生命是 ...t is ...fe?

[汉]对照(英)

Erwin Schrödinger

[奥] 埃尔温 · 薛定谔 著

杨焕明 译

CTS K 湖南科学技术出版社
·长沙·

Based on lectures delivered under the auspices of the Dublin Institute for Advanced Studies at Trinity College, Dublin, in February 1943

To

the memory of

My Parents

本书的撰写
基于本人1943年2月在都柏林圣三一学院高级研究院主办的讨论会上
所做的系列报告

谨以此书
纪念我的父母

目录

中英文对照版序（2023年）

研究生命的科学，即生物学，长期以来一直以一种隐含的理念为标志，即那些特定的原理使组成它的物体具有生命力。这种“活力论”在西方语言中始终非常明确，系统地倾向于将一个生命体的物质躯体与赋予其生命、精神或灵魂的非物质原理分开表述。这种思维方式可以追溯到很久以前，特别是它的初始阶段可以追溯到在希腊诞生的我们称之为科学的活动。在那里，第一原理被分为四类，即火、气、水、土。根据先知们的不同定义，它们被认为可以结合起来解释世界上的所有事物。

当然，在中国也有类似的原理，并多少有些异域情调：加入了第五种元素 —— 木，因为它把实实在在的生命引入了原始的自然科学之中。

但是，物理学之所以称为一门原创性学科（这正是古希腊语赋予自然研究的尊称），并不在于它所研究的对象，而在于苏格拉底之前的思想家们发明的独特方法，它使科学家（哲学家）逐渐以更充分的方式解释我们周围的世界。这种方式应该成为科学的永久根基，但却往往被忽视。要知道事实并不会说话，仅仅通过收集数据不可能取得进展。因此，要理解事实，必须采用不同于积累事实的方式。正是生物学在20世纪初突然进入了真正的科学世界，解释了为什么正是理论物理学才是现代生物学的源头，因而取得了最惊人的进展。

希腊先贤的基本思想是：我们必须成为发现世界的积极行动者，而不是单纯的袖手旁观者。知识的进步既不是由神灵或祭司引领，也不是靠任何直觉的驱动。科学不是通过什么启示就能揭开的秘密，而是一种人类的构建，一种即时性的构建。但这种即时性有其特殊之处，在于它不会受到任何方式的限制，它不能依靠墨守成规，而是不断前进。它通过在先前认识的基础上增加新的发现来进行，而不是无缘无故地拒绝或提出任何奇思妙想。

因此，（古希腊的）科洛丰城的哲学家色诺芬尼（Xenophanes）就指出，科学是建立在深思熟虑的假设之上的，并且这一过程与哲学大师们喜爱的那些谜语的制定密不可分：“至于某些真理，没有人看到过，正如没有人会真正了解关于神或我所说的任何事情。即使他完全成功地说出了真相，他也无法了解它，因为人们只能对所有事物进行猜测。”

由此可见，学者的工作目标是构建一套连贯的命题，以定律的形式组合起来，支配所有的知识领域，而不是发现一个隐藏的秘密。这确实是人类思想的产物，虽然不是这个或那个什么启示。这种方法包括从临时接受的假设中建立一个模型，以测试其对现实的适应性 —— 以存在性预测的形式，但更多的是通过模型预测的证伪来完成。

正是这种做事方式，解释了薛定谔（Schrödinger）在1944年出版的这本小书《生命是什么？》中描述的活细胞的物理性质，作者当时在都柏林流亡以逃避纳粹的迫害。

这项工作有什么新颖、革命性的意义呢？80年后，我们可以理解薛定谔思维的前瞻性以及对研究对象的适用性，但也可以看出他忽略了生物学的核心问题 —— 即生命化学表现出来的“活力”现象的物理基础。这可能解释了为什么虽然物理学家确实是现代生物学，早期被称为“分子生物学”的创始者，但相关的概念性发现在DNA复制、遗传密码表和基因表达或异构调节的概念性发现之后就停滞不前了。尽管薛定谔有着宏伟的雄心，但物理学并没有能够确定属于物理学的真正规律，其对生命的表达是非常特殊的。这是因为构成生物体基底的化学的表面“活力”现象，对于大多数人来说，包括物理学家在内，仍然是一个谜。

阅读《生命是什么？》使我们能够理解这一点：除了对生物学很重要的几个物理学概念以外，尤其是信息实实在在成为无处不灵的“硬通货”，我们还发现在这本书中，如果不是源头，那么第一个真正强调的主题是支配讨论并将古老的四元素理论的神秘魔力偷偷地重新引入到生物学本身。这种魔力来自于存在一个神秘的潜在、非形式化的吸引人的原则的隐藏假设，使得那些聚集的物体可以自发地有条不紊地组织起来。实际上，一种反复出现的思想困扰着当时的一些想法 —— 热爱生物学但没有真正勇气或愿望去理解生命，试图滥用一种以更具社会政治性而非物理本质的意识形态来概括、唤起基于“自我组织”的非常原始和非常神秘的想法的“秩序”来描述整个世界。

我们本可以从物理学中期待更多。那么，为什么现代物理学基本上只为生物学带来了尖端技术，尽管很多很多，而源于物理学的创新概念却如此之少？遗传密码的规则将多肽（蛋白质）与核酸准确无误地对应起来，这个规则无法从薛定谔方程中推导出来，尽管它当然与其完全兼容。它是如何出现在“普通”化学中的呢？这难道不是我们应该理解的吗？这意味着要理解一个有关区分不同类别对象的明确概念的物理基础，犹如理解核糖体如何通过选择正确的tRNA而准确地识别密码子-反密码子配对。

非周期性晶体

早在1935年，德国物理学家马克斯·德尔布吕克（Max Delbrück）作为一个天真的物理学家（据薛定谔的用语），就试图探究生命的物理化学基础。问题是尝试将遗传学的形式和非常抽象的关于某些遗传性状的传递的预测性的遗传学定律与细胞的组成部分，特别是细胞核和染色体的物理性质联系起来。这种思考的起点是对突变的研究，突变被理解为在一个个体的后代中发生的突然、不连续却颇为稳定的变化。离子辐射（如X

射线）对细胞的作用使得德尔布吕克和他的同事们能够具体地将遗传单位（基因）与细胞中离子辐射的主要目标（靶）具体联系起来。这样的计算结果表明，这个靶非常小，只有分子的大小范围，最多只是由几千个原子组成。这是薛定谔思考的起点，与认识论的预设有关，可以用以下问题和陈述来概括：

“在一个生命体的空间界面上发生的时（间）空（间）事件，如何用物理学和化学来解释？”

薛定谔表明，孤立的原子或小分子将无法保存准确传递遗传性状所需的记忆。这给他提供了一个机会，可以将他所知道的物理学与原子物理学联系起来，并将统计力学强加于真正的生物事实之上。特别是，他长时间地专注于生命的一个基本属性的思考，即所有生命都能以全方位的、宏观的方式行事。尽管这非常重要，但今天经常被忽视。

如何将原子的聚集与单个细胞的运动与人脑的组织联系起来？所有的问题都在于找出这种聚集的方式和结构：只有原子的组合才能做到这一点，前提是这一组合是足够稳定的，在生命发育的温度条件下具有足够长的寿命。当大量基本元素聚集在一起时，微观和宏观之间的对话就可以发生了，如顺磁性、布朗运动或扩散等简单的例子。我们的作者证明了可能存在空间和时间的宏观秩序，并且适当的局部和全局动态之间的来回运动可以创造一个稳定的宏观形式。例如，人们在那里发现了某些对分子生物学非常苛刻的思想家的所有反思，如大名鼎鼎的数学家勒内·托姆（René Thom）。今天，再度重视生命的这一重要维度，即宏观和微观之间的耦合，无疑是十分有益的。但是，正如我们将看到的，还有其他属性，不涉及那么多的物理条文，却解释了细胞行为的全部特征。

薛定谔的杰出贡献（与一些批评家所说的相反，他们被生物学家的理论缺陷所困扰，忽略了在所考虑的对象中选择最小尺度的相关性）在于揭示了分子水平，无论多么微小，都可以被视为一个相关的分析层次，甚至用于遗传性的组织。这就是现在被称为分子生物学先驱的物理学家，如威尔金斯（Wilkins）或弗朗西斯·克里克（Francis Crick）阅读薛定谔的文章时着迷的原因。我们通常对物理学或化学的印象是在考虑有组织的结构时，原子的排列是有规律的（这是晶体学的基础）。晶体是无机世界的模式，也是这个物质世界服从物理定律的象征。

生命世界的特点是不规则并千变万化的运动。这个世界的物质具有流动、粘连、黏稠的物质特性。事实上，当时的化学家把从生物系统中提取的产物称为胶体。薛定谔假定掌管物质塑造的法则具有绝对一致性，无论是惰性的还是活性的。他必须调和这两种极端的特性，即无机晶体和有机非晶态。他需要另一种方法来达到他对生命物质中心的想法，即遗传的支持：无机原子的集合对热波动过于敏感，因此它们必须组成分子才能发挥这种作用。虽然他没有使用“大分子”这个在他的时代相对较新颖的、被大多数化学家使用的概念，但薛定谔认为遗传的原子按照晶体结构规定的顺序组织，但这种顺序不是简单有序的重复而可能导致惰性结构的贫乏，而是包含由晶体的规则矩阵承载的变动。这种非周期性晶体是遗传性记忆的支撑。这个比喻很容易理解，好奇的人们被它的力量所吸引，特别是它指引我们发现基因物质的真正组织方式，这一点并不令人惊讶：

“我们相信一个基因——或许整个染色体纤丝——就是一个非周期性的固体。”

作为注解，有一句子具体指出要看什么：

“毫无疑问染色体纤丝非常柔韧，就像一根纤细的铜丝那样。”

道路已经铺平：将染色体的物质材料纯化，你就会发现非常纤细的丝状物（它的大小已经为德尔布吕克的X射线诱变实验所阐明），其结构具有非周期性晶体的特征！

同时，艾弗里（Avery）和他的同事发现了DNA的转化能力：基因的化学性质被发现。在《生命是什么？》出版后不到十年的时间里，那些曾经读过薛定谔的小书并受其启发的研究者们揭示了这种非周期性晶体的本质。

编码信息的概念

我们这本书的作者还播下了许多其他发现的种子。源于他的思考和直观的想象，正如物理学中经常发生的那样，“思想实验”在其中起着主导作用，薛定谔试图强调研究生物体需要关注的问题。问题的相关性是至关重要的：它决定了答案的重要性。

薛定谔明白，仅仅知道遗传的介质是不够的，还必须了解这种介质的组织如何转化为与生命相关的行为：毕竟，一个生命体必须能够与环境进行物质交换，也许还要乔迁搬家，肯定还要繁衍后代。在那个时候计算机科学尚未出现，而“程序”一词则隐含于戏剧相关的节目单里。薛定谔使用了一个比喻：为了描述这一转变以及和基因基质的作用，他使用了今天很多计算机科学家的行话——“密码”一词（但是以“密码本”的形式，在今天我们可能使用“程序”这一词语），这是一种以严格的决定性方式决定一个生物体未来发展的程序：

“把染色体纤丝结构称为密码本，我们在这里是指拉普拉斯（Laplace）曾经设想过的能洞察未来的意念，它对每一种因果联系都能立即阐明，它可以从卵的结构中看出，在适宜的条件下，这个卵将发育成一只黑公鸡还是一只芦花母鸡，是长成一只苍蝇还是一棵玉米植株，是一株杜鹃花还是一只甲虫，是一只老鼠又或是一位女士。”

这正是一种在当时并不广泛使用的编码概念，只用于秘密情报的交换。薛定谔将之比喻为摩尔斯（Morse）电码，只需要极少的基本元素就能够创造出无限数量的有意义组合。这是一项基本的生物学定律，直至今日仍然经常被一般大众所误解。而薛定谔的观点的核心，只有他的一些读者能够理解，也就是有机体构建的模式，即所谓布丰（Buffon）的“内模”（布丰是18世纪法国最重要的博物学家之一，在他的《自然史》一书中就提出了几个描述生命的重要概念，如每一种动物和植物都是由“有机分子”构成。每个生物体有机分子都带有每个物种特有的“内模”标记。在生殖过程中，雄性和雌性产生一些带有亲本“内模”标记的有机分子，供后代继承——译者注），这是一个物理的东西，由它自定的规则来操纵。但此时薛定谔走了一条错误的捷径，认为非周期性晶体本身是构建细胞中的一个参与者。

“当然，‘密码本’一词太过于简单了。因为染色体结构同时也负责引导卵细胞按照它们的指令发育。也就是说，染色体是法律条文与执行能力的统一，打个比方，它集设计师的蓝图与建筑者的技艺于一身。”

或者，

“这也不是随便什么密码都能被采用的问题，因为密码本其本身还必定是发育过程中的操纵因子。”

然而，这句话引起了误解，因为它似乎让人误以为非周期性晶体具有构建者的技能。正如我们今天所知，构建者的技能对应于基因的表达，而不是基因本身。而且薛定谔无法明确地表达这个观察结果。必须构想一个实现机制的程序，这个机制必须被确定。此外，密码本和构建者技能/程序之间的混淆无疑给遗传密码的规则的发现和理解，特别是对公众对它的理解带来了障碍。这个规则允许蛋白质的一个氨基酸通过RNA与基因中存在的核苷酸序列相对应，正如克里克的适应体（adaptor）概念所理解的那样，所有更新的规则都被薛定谔忽略了。

基因表达的准确性

如何理解程序的持久性，以及生物一代复一代的稳定性？即使这个问题看起来很棘手，量子物理学家薛定谔也没有回避。他认为基因是一个多原子结构，他相信他必须将通常适用于生命发生的温度下原子命运的规则应用于基因。因此，他试图用能级和这些级别之间的“量子跃迁”来描述基因的“非周期性”特征。

值得一提的是，我们应该注意到在这里薛定谔是多么的清醒，因为我们希望今天的许多“量子”物理学家也能像他一样，仍然知道“量子跃迁”是量子模型不连续性质的悖论之谜。所有的问题都在于支持遗传的分子的稳定性，以及它繁殖的精度，因为它必须从一代到下一代进行复制。

薛定谔表明，分子是一个多原子结构，这个概念本身就可以对生物体的时间尺度施加足够的稳定性约束。真正的问题是还必须考虑突变，从同一模式可能出现非常多的不同结构。然后薛定谔想象了一个典型大分子的异构体，但是他的描述却止步于此。他错过了语言比喻的核心思想：基因的物质支持是由四种基本单位的线性序列组成，其顺序决定了基因产物的性质。一千个这样的基本单位允许出现如此之多的组合，以至于任何空间结构在原则上都能够在其中找到一种表示方式。但是如何从基因得到它所决定的内容？以及如何确保其产物随着时间的推移而不会发生变化？

尽管存在不完美之处是必然的——必须承认，薛定谔当时只能通过借用德尔布吕克的实验和模型提出或多或少无根据的假设——但他的文章中包含了一种基本反思的萌芽。有趣的是，这种反思在30年之后才被美国物理学家约翰·霍普菲尔德（John Hopfield）提出，并且直到我们自己对能量耗散、基因表达的准确性、以及生物体类别的判别进行研究时才被提出。他确实表明，遗传支持的稳定性是至关重要的，热运动的影响是引入一定概率的“自发”基因性质变化。这些变化在有限的时间后发生，取决于支持“非周期性晶体”的分子性质，并且它们必然具有预期的突变的不连续性质。

遗传在时间进程中的一致性因此引发了问题。薛定谔提出了一种基于量子物理学的特别推理，以由于量子跃迁而产生的可能阈值为基础，从而避免热波动的统一后果，但他并没有进一步推理：如果突变是不可避免的，是否有一种方法来纠正它们或者至少在生物繁殖时控制它们？精确度问题是一个紧迫的问题：今天的语言比喻非常适合此问题。将基因的表达比喻为将其转写为RNA，然后进一步转写为蛋白质，但这在1944年并不存在。如何确保一本书的最终印刷没有排版错误？如何正确地校对文本以实现有效的修正？这样的问题对物种的稳定性非常关键。薛定谔的推理模型应该是有用的，它仍然有助于我们强调物理约束对生物繁殖的精度和准确性的重要性。有待发现的机制的前提是，细胞能够区分对象的类别，即对象的相同和变异，并找到尽可能消除变异体的方法。这个概念完全超出了我们的作者的思考范围。

薛定谔是时间意识形态的受害者，他想象生命是一场对抗熵的斗争。

这个非凡观点是如何被扭曲的，导致绝大多数物理学家不可挽回地走上了错误的道路？出发点是毫无疑问的，也没有错误：

“生命活动遵循物理学定律”，

但很快它被改成了一个可疑的形式，不是在细节上，而是在滥用的概括上：

“基于原子统计学的物理学定律只是近似的”，

当然，这是正确的，但前提是要避免紧随其后的反思，将物理学从程序的概念及其表达中移除：

“物理定律的精确度需要巨量原子的参与。”

第一个例子是在顺磁作用过程中产生的形式，无疑是对1925年由伊辛（Ising）提出并在1944年非常流行的模型的暗指。问题的确不在于由大量的数据所获取的精确度——实际上，许多细胞中存在的物质的数量级很小——而在于对信息本质的反思，作为真实世界的一个真正类别，以及保持细胞所产生的物质质量的可靠途径。这就需要一种新的物理学方法，基于区分不同类别的相似或不相似的物质（例如青年人和老年人），这个概念在当时是缺失的，而在当代物理学中仍然几乎完全被忽略。

此外，薛定谔偏离了他的第一次思考，即他正确地确定了分子水平，不仅从简单的细胞来分析遗传，而且，通过不幸的普遍的人类中心主义，对多细胞生物，特别是动物特别感兴趣。最后，他又回到了科学不存在的“奇妙”世界，他甚至写道：遗传、基因型和表型之间的联系可能难以理解：

“真是个奇迹——我的意思是：我们的全部存在，完全是依靠这个奇迹的这种相互作用，而我们又好像有能力去获得有关这种奇迹的更多知识。我认为，我们是有可能掌握这些知识的，并且这些知识可以引领我们进一步接近第一个奇迹。第二个可能超越了人类的认知范围。”

另一方面，通过非周期性晶体的比喻，我们可以理解突变：

“我们应该这样设想，基因结构本身就是一个巨大的分子，能够进行不连续的改变，即原子的重排，并组成一个同分异构的分子。这一重排也许只影响这个基因的一个小小的区域，而且可能发生了大量不同的重排。将实际的构型从任何可能的同分异构分子中区别开来的能量阈值一定要足够大（与一个原子的平均热能相比），大到足够将之转化为一个罕见事件，这些罕见事件就是我们所说的自发突变。

本章的后面部分将致力于对基因和突变的概貌（主要是德国物理学家德尔布吕克的模型）进行验证，通过将其与遗传学证据进行详细比较。在此之前，我们还是先对该理论的基础和一般性质作一些说明。”

到目前为止，薛定谔对物理学和生物学发展的贡献是非常正面的。不幸的是，他所生活的时代无疑导致了他思想的扭曲，一种认为世界不可避免地退化，需要与无序作斗争的意识形态在当时占主导地位。薛定谔受到很多物理学的统计思想的影响，忘记了细胞远非总是由大量有意义的物体组成，将其与不可避免的无序思想联系在一起。他将这一观点与一个常见的但确是错误的关于熵的概念联系在一起。他定义了熵和无序，并援引热力学第二定律作为动力，生命必须不断地以“负熵”为食，并与之斗争。

然而，薛定谔在一篇长文中谨慎地指出，他的物理学同事并不同意他的观点：

“关于负熵的说法遭到了物理学同行们的质疑和反对……西蒙(Simon)非常正确地向我指出：事实上，我们的食物中所含的能量确实很重要。因而我对菜单上标出食物所含能量的嘲讽并不合适。我们不仅需要能量来提供身体活动所需的机械能量，而且需要它补充由身体不断释放到环境中的热量。我们散发热量并不是偶然的，而是必不可少的。因为我们正是以这种方式来清除在生命活动过程中不断产生的多余的熵。

这似乎表明，温血动物拥有较高的体温这一优势，能以较快的速度排出身体产生的熵，因此能够承受强度更大的生命活动。我并不能肯定这个观点在多大程度上符合实情(对此负责的是我，而不是西蒙)。有人可能会反对这种观点，因为另一方面，有许多温血动物以皮毛或羽毛来防止热的快速散失。”

确确实实，熵必然增加，但这并不自动意味着所涉及的对象的无序性增加。无序的概念已经部分地出现在玻尔兹曼(Boltzmann)关于统计力学的文章中。但相关的“证明”并非真正的证明，而仅是对世界的某种观察。为什么我们需要这两个词，无序和熵，来表达同样的事情呢?相反，难道我们不能从物质的自发行为中看到探索和创新的能力吗?占据所有可用空间的事实允许事物在其初始空间前面进行不可预见的相互作用：通过混合黄色和蓝色创造绿色，是否存在无序?探索倾向于无序还是有序，取决于环境。事实上，我们今天知道，正是由于热力学第二定律的作用，它强制任何物质系统的熵自发增加，才创造了定义生命的诸多重要形式的大分子。溶解在二甲基甲酰胺中的DNA分子构成一组随机折叠的螺旋，没有特定的形状，但是一旦我们加水稀释，就形成了一个双螺旋。的确存在一种有序的创造，与熵的增加并行，而这种有序是由水分子的重新排列驱动的：混合物的熵增加趋势是通过强制外来分子采用特定的构象，尽可能地为水分子留出空间。但是，薛定谔关于他的同事西蒙的评论也引发了一个直到最近才被理解的问题。正如我们将在结论中看到的那样，在与生命相关的能量耗散中确确实实有一些东西需要理解。

像几乎所有想要概括其知识的作者一样，薛定谔发现自己被他的冲动冲昏了头脑。他想要在他的思考中赋予普遍的维度，显然这带来了风险。量子模型如此强大，当它证明遗传的支持必然具有分子多原子的性质时，难道也可以通过统计力学来考虑形式在生命中的作用吗？这里显然提出了生命现象中组织、秩序和无序的问题。

但是，1944年的意识形态是否有利于这种思考呢？纳粹主义曾试图统治世界，在薛定谔写书时，纳粹主义仍然野心勃勃。因此，在一个遗传学占据核心地位的思考中，发现一些非常有争议的元素，并且薛定谔也意识到了这一点，这并不奇怪。这种情况的第一个证据是有关分子扩散的“肘部空间”（让人想起“Lebensraum”，希特勒声称为德国人要求的“生存空间”）一词和概念的奇怪引入：当分子扩散时，真的是为了有更多的生存空间吗？当然，与当时的背景最相关的是突变：

“说得严重点，虽然可能有点天真，但堂亲、表亲之间的婚配带来的伤害的程度也可能因为他（她）的祖母曾长时间从事X射线护士的工作而增加。虽然，这不是任何个人需要担心的问题，但任何慢慢影响人类的不理想的潜在突变都应当受到社会关注。”

或具有明显的人道主义立场：

“既然如今我们不再用斯巴达人在泰杰托斯山经常采用的那种残暴方式去消灭弱势者，那么，在人类中，自然的适者生存的选择作用大大减弱了，不！是径直走向对立面，我们必须严肃地看待在人类中发生的这些事情。在更原始的条件下，战争也许对选择出最合适生存的部落有积极的价值；而现代大量屠杀所有国家的健康青年的反选择效应，连这一点理由也不存在了。”

这是薛定谔作为先驱者的另一种表现：这难道不是超前的社会生物学吗？然而，我们的作者非常清楚突变的偶然性。他对德尔布吕克工作的分析在这方面毫不含糊：

“就目前的情况而言，我们可以得出结论：假如一半辐射的剂量导致了，比如说，千分之一的子代个体发生了突变，但是丝毫不影响其余的子代，无论是在使他们易于发生突变的方面，还是在使他们产生对突变的免疫能力方面。否则，另一半的辐射剂量就不能正好诱发千分之一的突变体。因此，突变并不是一个累积效应，不是由连续的小剂量辐射相互增强而累积引起的。”

因此，与同时期的遗传学家如杜布赞斯基（Dobzhansky）一样，薛定谔重新解释了达尔文，考虑到孟德尔的定律和德弗里斯（De Vries）关于突变的研究。从这个意义上说，他是新达尔文主义潮流的缔造者之一，将主导分子生物学的诞生。他和当时的许多思想家犯了同样的错误（现今仍然普遍存在），即没有清晰地区分什么是表现型和什么是基因型，也没有理解遗传的两个基本方面，即其基因特征（DNA的精确复制）和表观遗传特征（DNA表达一次的准确产物）。但他非常清楚，即使在今天，很少有人理解选择基于对突变的先前选择，没有任何环境的指导性影响。

毫无疑问，他知道1943年卢里亚（Luria）和德尔布吕克进行的精妙实验，这些实验证明了突变的纯粹偶然性，而不是任何选择。但是，正如作者自己承认的那样，他写这本书的唯一目的是要表明，除了物理定律之外，还有其他的定律，这些定律在不违反物理定律的前提下，通过生命科学得到特别的展现。他保留的想法很符合当时的意识形态，即世界自然地趋向于无序，而生命体特有的定律的作用是不断地抵制这种自然倾向。由于薛定谔论证的后果是非常值得商榷的，今天仍然存在于某些知识界的老生常谈中的许多评论中，因此有必要稍微详细地回顾一下它们。

初步的结论

这是一个基本问题：生命体的有序组织是如何维持的？我

们在前面已经看到，在生命的延续中，预测大分子结构的复制是核心问题。但是，要使一个细胞持续存在，甚至不考虑其繁殖问题，随着时间的推移，还必须使它的元素尽可能地接近常态，并保持其空间组织。面对生命结构的明显复杂性，这似乎是神奇的，现在看来依然如此。人们可以理解为什么许多人诉诸出现活力论暗示的观点。在19世纪热力学诞生之初，一些人提出了能量这个模糊概念，其中可以插入“生命能量”，这种观念在今天面向那些受教育程度较低的人群的广告中仍然存在。但是，由于熵这个神秘而难以定义的概念，类似的误导也出现在教育程度较高的圈子里。今天，活力论远未消亡，即使它被掩盖在“自我组织”的名称之下，这个伪概念基于帕斯卡尔·乔丹（Pascual Jordan）在20世纪30年代提出的同类实体之间“互补”的错误观点，并在1940年被德尔布吕克和鲍林（Pauling）驳斥。

正如我们之前所看到的，物理学及其定律确实扮演了至关重要的角色。正是物理学家罗尔夫·兰道尔（Rolf Landauer）在1961年的工作（薛定谔不太可能知道），提供了人们理解生物化学活力的重要线索。这包括两个关键步骤：

第一步是“约束”，这个步骤包含了信息，涉及到能量来源的捕获，但尚未消散能量，并保留了信息的量子状态（通常是在包含相邻类型分子的环境中，通过选择一个特定的分子来实现，例如一个相对于其年轻的对应物而言的老化分子）。

第二步是重置（或“松弛”），在这一过程中，系统释放能量以将其恢复到基本状态，以允许进程重新启动。

目前在最小的合成基因组中发现的这类功能，其作用类似于麦克斯韦（Maxwell）在其1872年的《热力学理论》中提出的那个东西，今天被称为“麦克斯韦妖（Maxwell Demon）”。它可以在其他类似底物中区分底物，识别空间结构中的一个特定位置，或一组连续事件中的一个特定瞬间。

这为我们带来了一个充满希望的信息。迟做总比不做好：在《生命是什么？》出版不到一个世纪之后，薛定谔的梦想是找出生物体独特属性背后的物理定律，这即将成为现实。

安东 · 唐善（Antoine Danchin）
法兰西科学院院士
法兰西科学院分子细胞生物学和基因组学学部外事代表
2023年3月8日

英文再版序（1991年）

20世纪50年代初，当我还是一名年轻的数学系学生时，并没有读过很多书，但我读过的那些——至少从头到尾读过的书——通常就是埃尔温·薛定谔的作品。我不得不说他的作品总是那么引人入胜，给人一种发现的崭新喜悦，激发了我们对生活的这个神秘世界真正的新的理解的期待。在他的所有作品中，没有一本书能达到他的经典小书《生命是什么？》这样的高度。据我所知，这一定是21世纪最有影响力的科学著作之一。它代表了一位物理学家理解一些生命的真正奥秘的一次强有力的尝试，他自己的深刻见解极大地改变了我们理解生命世界构成的方式。这本书的跨学科之广在当时是不同寻常的——然而它以一种亲近可爱的，让人放松的，谦虚的笔法写成，让非专业人士和有志于成为科学家的年轻人都能读懂。确实，许多在生物学上做出重大贡献的科学家，如霍尔丹和弗朗西斯·克里克，都承认深受（尽管并非完全赞成）这位极富原创性和思维缜密的物理学家所提出的广泛思想的影响。

就像许多对人类思维产生重大影响的著作一样，这一小书所提出的观点一旦被人们掌握，就是一连串几乎不言而喻的真理，然而，这些观点仍被相当大比例的人视而不见，而这些人本应更多了解相关情况。我们是否还经常听到量子效应与生物学研究几乎没有关联，甚至我们摄食只是为了获得能量？

这就需要再度强调薛定谔的《生命是什么？》持续至今的重要意义，它非常值得我们再次阅读！

罗杰·彭罗斯（Roger Penrose）
1991年8月8日

英文首版作者原序（1944年）

一个科学家通常被认为应该对某些学科有着完整且深入的第一手知识，因此，通常不期望他在自己不精通的领域里著书立说。这被认为是所谓的“贵族义务”。但是为了把这本书写好，我恳求让我暂且放弃“贵族”这一贵冠，继而免除这一义务。我的托词如下：

我们从祖先那里继承了对统一的、包罗万象的知识体系的热切渴望。最高学府之所以称为最高学府（即大学），就是为了警醒我们，从古至今，多少个世纪以来，“普适性”才是唯一推崇的信条。但是，在过去的百年中，五花八门的知识分支不管是从其广度还是深度来说，都给我们以新的困惑。我们清晰地感受到，我们现在才刚刚开始获得可靠的材料，并将其汇总成为一个有机的整体。但是，从另一方面来讲，让一个人完全掌握超过某一领域的专门知识几乎是不可能的。

我已看到这个困惑是难以逾越的（为了我们不至于永远放弃追求真理的目的），只好我们之中的一些人鼓足勇气冒险上阵，冒的是愚弄自己的风险，目的是要将一些事实和理论归纳成一个整体，尽管其中一些可能不是第一手的支离破碎的知识。

这就是我的致歉。

另外，语言方面的障碍也是不容小觑的。一个人的母语就像一件合身的外衣，如果不能信手拈来，你就永远不会满意，只能换上另一件。我要感谢英克斯特博士（都柏林的圣三一学院），帕德里克·布朗博士（梅诺斯的圣帕里克学院），最后还有做出同样贡献的S. C. 罗伯特先生。他们竭尽全力地让我的新衣“合身”，而我偶尔固执地选用那些“原创”时装，更给他们增添了诸多麻烦。如果书中还有一些“原创”由于我朋友的宽容而存在，责任在我，而与他们无关。

很多节段的标题都来自我原先旁注的归纳，每一章的正文也许需要连贯前后来阅读。

埃尔温·薛定谔
1944年9月于都柏林

自由之士绝少想到死亡，他的才智是对生命的沉思，而不是对死亡的冥想。

——斯宾诺莎《伦理学》第四卷，命题67

CHAPTER 1

The Classical Physicist's Approach to the Subject

Cogito ergo sum. DESCARTES

THE GENERAL CHARACTER AND THE PURPOSE OF THE INVESTIGATION

3 This little book arose from a course of public lectures, delivered by a theoretical physicist to an audience of about four hundred which did not substantially dwindle, though warned at the outset that the subject-matter was a difficult one and that the lectures could not be termed popular, even though the physicist's most dreaded weapon, mathematical deduction, would hardly be utilized. The reason for this was not that the subject was simple enough to be explained without mathematics, but rather that it was much too involved to be fully accessible to mathematics. Another feature which at least induced a semblance of popularity was the lecturer's intention to make clear the fundamental idea, which hovers between biology and physics, to both the physicist and the biologist.

第一章
经典物理学家对生命的探索

我思故我在。

——笛卡尔

生命奥秘之探索

这本小书源于一场理论物理学家所做的公众演讲，400余名听众几乎全程没有离席，虽然一开场时他们就被告知这一学科晦涩难懂并且这个演讲谈不上多受欢迎，即使物理学家那最令人生畏的“武器”——数学推导也很少使用。其原因并不是这个主题简单到不用数学推导就可以讲清楚，相反，正因为它太复杂，无法完全用数学来解答。这一演讲的另一个至少在表面上受欢迎的特点是：演讲者力图阐述清楚那些徘徊于生物学和物理学之间的基本概念，以便同时让物理学家和生物学家都能理解。

For actually, in spite of the variety of topics involved, the whole enterprise is intended to convey one idea only—one small comment on a large and important question. In order not to lose our way, it may be useful to outline the plan very briefly in advance.

The large and important and very much discussed question is:

How can the events *in space and time* which take place within the spatial boundary of a living organism be accounted for by physics and chemistry?

4 The preliminary answer which this little book will endeavour to expound and establish can be summarized as follows:

The obvious inability of present-day physics and chemistry to account for such events is no reason at all for doubting that they can be accounted for by those sciences.

实际上，尽管涉及的话题很不一样，但整个演讲只是为了阐述一个想法——对一个重大问题的小小的评论。为了避免离题甚远，预先扼要地梳理一下计划也许是必要的。

一个重大的，且正被广泛讨论的问题是：

在一个生命体的空间界面上发生的时（间）空（间）事件，如何用物理学和化学来解释？

这本小书着力探讨和认可的初步答案可以归纳如下：

虽然这些生命活动还难以用今天的物理学和化学来解释，但是我们没有理由质疑它们的确能用这些科学理论来解释。

STATISTICAL PHYSICS. THE FUNDAMENTAL DIFFERENCE IN STRUCTURE

That would be a very trivial remark if it were meant only to stimulate the hope of achieving in the future what has not been achieved in the past. But the meaning is very much more positive, viz. that the inability, up to the present moment, is amply accounted for.

Today, thanks to the ingenious work of biologists, mainly of geneticists, during the last thirty or forty years, enough is known about the actual material structure of organisms and about their functioning to state that, and to tell precisely why, present-day physics and chemistry could not possibly account for what happens in space and time within a living organism.

The arrangements of the atoms in the most vital parts of an organism and the interplay of these arrangements differ in a fundamental way from all those arrangements of atoms which physicists and chemists have hitherto made the object of their experimental and theoretical research. Yet the difference which I have just termed fundamental is of such a kind that it might easily appear slight to anyone except a physicist who is thoroughly imbued with the knowledge that the laws of physics and chemistry are statistical throughout.[1] For it is in relation to the statistical point of view that

1. This contention may appear a little too general. The discussion must be deferred to the end of this book, pp. 82-4.

统计物理学：结构上的根本差别

如果说我们只是试图激励那种未来才能成功解决至今未能解决的问题的希望，那就未免太平庸了。实际上，它有着更为积极的含义，那就是，它足以解释为何迄今为止我们对此问题仍然无能为力。

今天，下列进展应归功于生物学家，主要是遗传学家，在近三四十年来的创造性工作：对于有机体的真实物质结构及其功能已知之甚多且准确详明，个中缘由令现代物理学和化学仍然不能解释生命体体内在特定的空间和时间中到底发生了什么。

一个有机体最关键部分的原子排列及其相互作用方式，和迄今所有物理学家和化学家当作实验和理论研究对象的原子排列有着根本的差别。然而，我刚才所说的这种根本的差别对于任何人来说都是微不足道的，除非是一个物理学家，他完全了解物理学和化学的规律自始至终完全基于统计学[1]。因为从统计学相关的观点来说，生命体关键部分的结构，和我们物理学家或

1. 这个论点可能有点太笼统了，将在本书第七章的7-8节（原文第82-84页）再加讨论。

the structure of the vital parts of living organisms differs so entirely from that of any piece of matter that we physicists and chemists have
5 ever handled physically in our laboratories or mentally at our writing desks.[1] It is well-nigh unthinkable that the laws and regularities thus discovered should happen to apply immediately to the behaviour of systems which do not exhibit the structure on which those laws and regularities are based.

1. This point of view has been emphasized in two most inspiring papers by F. G. Donnan, *Scientia*, XXIV, no. 78 (1918), 10('La science physico-chimique décrit-elle d' une façon adéquate les phénomènes biologiques?'); *Smithsonian Report for 1929*, p. 309('The mystery of life').

化学家在实验室里或书桌边绞尽脑汁研究的物质完全不同[1]。几乎无法想象的是，这样发现的定律和规则应当可以立即应用到生命系统的行为上去，而这一系统却又不具有作为这些定律和规则基础的结构。

1. 道南在两篇极具启发性的论文中强调了这个观点。见《科学》24卷，78期（1918年），10页“物理化学能否描述生物学现象”；《史密森学会1929年年报》第309页“生命的奥秘”。

The non-physicist cannot be expected even to grasp—let alone to appreciate the relevance of—the difference in 'statistical structure' stated in terms so abstract as I have just used. To give the statement life and colour, let me anticipate what will be explained in much more detail later, namely, that the most essential part of a living cell—the chromosome fibre—may suitably be called *an aperiodic crystal*. In physics we have dealt hitherto only with *periodic crystals*. To a humble physicist's mind, these are very interesting and complicated objects; they constitute one of the most fascinating and complex material structures by which inanimate nature puzzles his wits. Yet, compared with the aperiodic crystal, they are rather plain and dull. The difference in structure is of the same kind as that between an ordinary wallpaper in which the same pattern is repeated again and again in regular periodicity and a masterpiece of embroidery, say a Raphael tapestry, which shows no dull repetition, but an elaborate, coherent, meaningful design traced by the great master.

In calling the periodic crystal one of the most complex objects of his research, I had in mind the physicist proper. Organic chemistry, indeed, in investigating more and more complicated molecules, has come very much nearer to that 'aperiodic crystal' which, in my opinion, is the material carrier of life. And therefore it is small wonder that the organic chemist has already made large and important contributions to the problem of life, whereas the physicist has made next to none.

不敢指望非物理学家能掌握我刚才用那么抽象的名词所表达的“统计学结构”的差异，更不用说去理解它的重要性了。为了让这一叙述更加生动有趣，让我先剧透一下后面将详细解释的内容，即一个活细胞最主要的组成部分——染色体纤丝——可以恰如其分地称之为非周期性晶体。在物理学中我们迄今研究的只是周期性晶体。

在一位谦逊的物理学家的脑子里，周期性晶体是十分有趣而相当复杂的研究对象。它们构成了非生命的自然界中最有魅力也最复杂的一类物质结构，使物理学家伤透脑筋。然而，与非周期性晶体相比，它们显得那么单调乏味。两者之间结构上的差别在于，周期性晶体就好似一张按照一定规律反复出现相同图案的墙纸；而非周期性晶体则是刺绣的杰作，就像一件拉斐尔[1]挂毯，它呈现的并不是乏味的重复，而是大师那精美绝伦、独具匠心的创作。

将周期性晶体称为研究中最复杂的对象之一时，我想到的是物理学家本身。有机化学研究的分子确实也越来越复杂，已经非常接近“非周期性晶体”了。在我看来，那些分子其实就是生命的物质载体。因此，有机化学家对生命问题已经做出了重大且重要的贡献，而物理学家迄今却几乎毫无建树也就不足为奇了。

1. Raphael，全名Rafaello Santi（1483—1520），意大利著名画家，代表了文艺复兴时期艺术家从事理想的事业所能达到的巅峰。

THE NAÏVE PHYSICIST'S APPROACH TO THE SUBJECT

6 After having thus indicated very briefly the general idea—or rather the ultimate scope—of our investigation, let me describe the line of attack.

I propose to develop first what you might call 'a naïve physicist's ideas about organisms', that is, the ideas which might arise in the mind of a physicist who, after having learnt his physics and, more especially, the statistical foundation of his science, begins to think about organisms and about the way they behave and function and who comes to ask himself conscientiously whether he, from what he has learnt, from the point of view of his comparatively simple and clear and humble science, can make any relevant contributions to the question.

It will turn out that he can. The next step must be to compare his theoretical anticipations with the biological facts. It will then turn out that — though on the whole his ideas seem quite sensible —they need to be appreciably amended. In this way we shall gradually approach the correct view—or, to put it more modestly, the one that I propose as the correct one.

Even if I should be right in this, I do not know whether my way of approach is really the best and simplest. But, in short, it was mine. The 'naïve physicist' was myself. And I could not find any better or clearer way towards the goal than my own crooked one.

天真的物理学家对生命的探索

非常扼要地交代了我们的主体思路，也可以说是研究的基本观点之后，现在让我来介绍我们的探讨路线。

首先，我打算进一步阐述也许被你称为“天真的物理学家的关于有机体的观点”，那就是一位物理学家学习了物理学，特别是形成他的科学的统计学基础之后，开始思考生命体及其行为方式和功能，他会认真地拷问自己：学到的知识，以及那些相对简单明了而又天真朴素的科学观点，是否有助于他为解决这些问题做出相应的贡献。

事实将证明他能。下一步必将是把理论预期和生物学事实相比较。结果证明，他的观点大体上是合理的，但需要做进一步的修正。这样，我们将逐渐接近这一正确的观点，或者谨慎一点来说，接近我认为正确的观点。

即使我在这方面是正确的，我也不知道我的这条探索之路是否真的是最佳和最便捷的。总而言之，这只是我的探索之路，这位“天真的物理学家”正是我自己。除了我自己这一条崎岖小路之外，我找不到通往这个目标更平坦更清晰的途径。

WHY ARE THE ATOMS SO SMALL?

A good method of developing ‘the naïve physicist’s ideas’ is to start from the odd, almost ludicrous, question: Why are atoms so small? To begin with, they are very small indeed. Every little piece of matter handled in everyday life contains an enormous number of them. Many examples have been devised to bring this fact home to an audience, none of them more impressive than the one used by 7 Lord Kelvin: Suppose that you could mark the molecules in a glass of water; then pour the contents of the glass into the ocean and stir the latter thoroughly so as to distribute the marked molecules uniformly throughout the seven seas; if then you took a glass of water anywhere out of the ocean, you would find in it about a hundred of your marked molecules.[1]

The actual sizes of atoms[2] lie between about $\frac{1}{5000}$ and $\frac{1}{2000}$ of the wave-length of yellow light. The comparison is significant, because the wave-length roughly indicates the dimensions of the smallest grain still recognizable in the microscope. Thus it will be seen that such a grain still contains thousands of millions of atoms.

Now, why are atoms so small?

1. You would not, of course, find exactly 100 (even if that were the exact result of the computation). You might find 88 or 95 or 107 or 112, but very improbably as few as 50 or as many as 150. A ‘deviation’ or ‘fluctuation’ is to be expected of the order of the square root of 100, i.e. 10. The statistician expresses this by stating that you would find 100 ± 10. This remark can be ignored for the moment, but will be referred to later, affording an example of the statistical $\sqrt{n}$ law.

2. According to present-day views an atom has no sharp boundary, so that ‘size’ of an atom is not a very well-defined conception. But we may identify it (or, if you please, replace it) by the distance between their centres in a solid or in a liquid — not, of course, in the gaseous state, where that distance is, under normal pressure and temperature, roughly ten times as great.

原子为何如此之小

要阐明“天真的物理学家的观点”的一个好办法，是以这个古怪可笑的、有点荒唐的问题作为起点：原子为何如此之小？

首先，原子的的确确很小很小。即使日常生活中我们使用的最小的一件物品，都含有数目巨大的原子。许多例子已经被设计出来以把这个事实展示给大家，但没有一个能比开尔文勋爵所用的这个例子更令人印象深刻：

假如你能够把一个玻璃杯中的所有水分子都做上标记，再把这杯子里的水倾倒入大海之中，然后充分搅拌海水，使得标记了的水分子均匀地分布到七大洋中；此后，如果你从海洋中任何一处舀一杯水上来，你会发现这杯水中大约有100个被标记上的分子[1]。

原子的实际大小[2]介于黄色光波长的1/5000到1/2000之间。这个比较是很有意义的，因为波长范围基本代表了显微镜可以辨认的最小颗粒的大小。而即使是这样的颗粒，也可以看到是由数亿个原子组成的。

现在可以问了，原子为什么如此之小？

---

1. 当然，你不会恰好在这杯水中就找到100个分子（即使这个结果是经过精确计算的）。你也许会找到88个、95个、107个或112个，但不大可能少于50个或多达150个。预期“偏差”或“偏移”是100的平方根，即10个。统计学家是这样表述的：你将找到100 ± 10个。这个注释现在可先暂略，后面还将提及。它是统计学的$\sqrt{n}$定律的一个实例。
2. 根据现代物理学的观点，一个原子并没有明确的界限，因此一个原子的“大小”并不是一个定义十分明确的概念。但是我们可根据固体或者液体中原子核的间距来确认。当然，在气体中这个理论是不成立的，因为在常温常压下，气态中的这个间距几乎要大10倍。

Clearly, the question is an evasion. For it is not really aimed at the size of the atoms. It is concerned with the size of organisms, more particularly with the size of our own corporeal selves. Indeed, the atom is small, when referred to our civic unit of length, say the yard or the metre. In atomic physics one is accustomed to use the so-called Ångström (abbr. Å), which is the 10^{10}th part of a metre, or in decimal notation 0.0000000001 metre. Atomic diameters range between 1 and 2Å. Now those civic units (in relation to which the atoms are so small) are closely related to the size of our bodies. There is a story tracing the yard back to the humour of an English king whom his councillors asked what unit to adopt—and he stretched out his arm sideways and said: 'Take the distance from the middle of my chest to my fingertips, that will do all right.' True or not, the story is significant for our purpose. The king would naturally indicate a 8 length comparable with that of his own body, knowing that anything else would be very inconvenient. With all his predilection (偏好) for the Ångström unit, the physicist prefers to be told that his new suit will require six and a half yards of tweed—rather than sixty-five thousand millions of Ångströms of tweed.

It thus being settled that our question really aims at the ratio of two lengths—that of our body and that of the atom—with an incontestable priority of independent existence on the side of the atom, the question truly reads: Why must our bodies be so large compared with the atom?

很明显，这个问题只是一个避而不谈的借口，因为真正的问题并不在于原子的大小，而是生命体的大小，尤其是我们自己身（肉）体的大小。的确，当我们使用民用（常用）的单位，如码（1码约为0.9144米）或米来度量，原子很小很小。在原子物理学中，人们习惯用埃（简写为Å）来度量，这是1米的百亿分之一，或者用十进制表示为0.0000000001（10^{-10}）米。原子的直径范围在1～2埃。这些民用度量单位（与之相比，原子是如此之小）同我们身体的大小是密切相关的。

有一个追溯“码”起源的故事。这是一位英国国王的幽默故事，他的大臣问他要采用什么单位，他侧身伸出他的手臂，说：“量一量我的胸口到手指尖的距离，就行了。”不论这个故事是真是假，对我们来说都很有意义。这位国王自然而然地指定一个可以同自己的身体作对比的长度，知道用任何别的单位都将很不方便。不管物理学家如何偏爱“埃”这个单位，他还是喜欢别人告诉他这一新衣需用六码半（约为5.9436米）的花呢料子，而不是说成650亿个埃。

现在明确了我们提出问题的真正目的在于两个长度之比——我们的身体和原子的大小之比。首先考虑的是把原子作为一种特殊的、独立的存在。所以，这个问题应该真正理解为：与原子相比，我们的身体为何一定要这么巨大？

I can imagine that many a keen student of physics or chemistry may have deplored the fact that every one of our sense organs, forming a more or less substantial part of our body and hence (in view of the magnitude of the said ratio) being itself composed of innumerable atoms, is much too coarse to be affected by the impact of a single atom. We cannot see or feel or hear the single atoms. Our hypotheses with regard to them differ widely from the immediate findings of our gross sense organs and cannot be put to the test of direct inspection.

Must that be so? Is there an intrinsic reason for it? Can we trace back this state of affairs to some kind of first principle, in order to ascertain and to understand why nothing else is compatible with the very laws of Nature?

Now this, for once, is a problem which the physicist is able to clear up completely. The answer to all the queries is in the affirmative.

我能够想象，许多聪明的物理学或化学的学生也许会对如下事实深感惋惜：我们的每一个感觉器官，事实上或多或少构成了身体上极其重要的部分，而身体本来就是由无数原子组成的（考虑到原子与身体的比例），因此只能粗略估计，单原子的碰撞反应无法对其造成任何影响。我们看不见、摸不着也听不到单原子。我们假说中的原子远远不同于我们不那么敏感的感官所能够直接发现的物质，而且也无法通过直接观察来对这些原子进行检验。

一定是这样吗？是否存在任何内在的原因呢？我们能否从这一状况追溯到一些首要法则？进而探明和理解为什么世界上没有与大自然的客观规律相适应的东西呢？

现在，物理学家能够彻底搞清楚这个问题，而且所有问题都能得到肯定的答案。

THE WORKING OF AN ORGANISM REQUIRES EXACT PHYSICAL LAWS

If it were not so, if we were organisms so sensitive that a single atom, or even a few atoms, could make a perceptible impression on our senses—Heavens, what would life be like! To stress one point: an organism of that kind would most certainly not be capable of developing the kind of orderly thought which, after passing through a 9 long sequence of earlier stages, ultimately results in forming, among many other ideas, the idea of an atom.

Even though we select this one point, the following considerations would essentially apply also to the functioning of organs other than the brain and the sensorial system. Nevertheless, the one and only thing of paramount interest to us in ourselves is, that we feel and think and perceive. To the physiological process which is responsible for thought and sense all the others play an auxiliary part, at least from the human point of view, if not from that of purely objective biology. Moreover, it will greatly facilitate our task to choose for investigation the process which is closely accompanied by subjective events, even though we are ignorant of the true nature of this close parallelism. Indeed, in my view, it lies outside the range of natural science and very probably of human understanding altogether.

生命活动遵循物理学定律

如果不是这样，如果我们自己就是如此敏感的有机体，哪怕是一个原子，或那么几个原子，都能在我们的感知下留下可察觉的印象——天哪，那生命将会是什么样子？需要强调一点：那个样子的生命体肯定不可能发展出井井有条的思路。而这一思路在早期经历了漫长的一个又一个阶段，终于在诸多想法中形成了关于原子的概念。

尽管我们选择了上述观点，但是下述考虑也基本适用于除大脑和感觉系统之外的其他器官的功能。当然，对我们自身而言，唯一重要的是，我们的感觉、思考和认知。对于担负思维和感觉的生理过程来说，大脑和感觉系统以外的所有其他器官只是辅助的，至少从人类的角度来看是这样的，即使我们不是从纯客观的生物学观点来看问题。而且，这样的观点将极大地促使我们选择研究那种紧密伴随着主观事件发生的过程，尽管我们对这种紧密的平行现象的确切性质一无所知。事实上，在我看来，它超出了自然科学的范畴，而且很可能完全超出了人类所有的认知范围。

We are thus faced with the following question: Why should an organ like our brain, with the sensorial system attached to it, of necessity consist of an enormous number of atoms, in order that its physically changing state should be in close and intimate correspondence with a highly developed thought? On what grounds is the latter task of the said organ incompatible with being, as a whole or in some of its peripheral parts which interact directly with the environment, a mechanism sufficiently refined and sensitive to respond to and register the impact of a single atom from outside?

The reason for this is, that what we call thought (1) is itself an orderly thing, and (2) can only be applied to material, i.e. to perceptions or experiences, which have a certain degree of orderliness. This has two consequences. First, a physical organization, to be in close correspondence with thought (as my brain is with my thought) must be a very well-ordered organization, and that means that the events that happen within it must obey strict physical laws, at least to a very high degree of accuracy. Secondly, the physical impressions made upon that physically well-organized system by other bodies from outside, obviously correspond to the perception and experience of the corresponding thought, forming its material, as I have called it. Therefore, the physical interactions between our system and others must, as a rule, themselves possess a certain degree of physical orderliness, that is to say, they too must obey strict physical laws to a certain degree of accuracy.

于是我们又将面临下一个问题：为什么像我们大脑这样的器官以及其连接的感觉系统，一定也是由不计其数的原子构成，才能让它的物理学变化状态与高度发达的思想密切联系呢？在什么基础上，我们所说器官的上述任务，作为与环境直接相互作用的全部或部分组成器官，是与下述机制相矛盾的呢？这一机制，是以一个足够精细和敏感的机制，来响应并记录来自外界的单原子的影响。

理由是这样的，我们所说的思想：（1）它本身是一个有序的东西；（2）只能应用于具有一定程度的有序的物质，比如感知或经验。这就可以得出两个结果：第一，同思想密切对应的身体的物理性组织（如承载我思想的大脑）必须是十分有序的组织，这意味着在它内部发生的事件必须遵循严格的物理学定律，至少要达到非常高的精确度；第二，外界其他物体对那个物理学上组织得很好的系统所产生的物理学作用，显然与相应的思想的感知或经验是一致的，这就构成了我所说的思想的物质。因此，我们身体的系统与其他系统之间一定存在物理学上的相互作用，并且作用本身具备了一定程度的物理有序性，也就是说作为一种规律，它们也必须在某一特定的精确度上，严格遵守物理学定律。

PHYSICAL LAWS REST ON ATOMIC STATISTICS AND ARE THEREFORE ONLY APPROXIMATE

And why could all this not be fulfilled in the case of an organism composed of a moderate number of atoms only and sensitive already to the impact of one or a few atoms only?

Because we know all atoms to perform all the time a completely disorderly heat motion, which, so to speak, opposes itself to their orderly behaviour and does not allow the events that happen between a small number of atoms to enrol themselves according to any recognizable laws. Only in the co-operation of an enormously large number of atoms do statistical laws begin to operate and control the behaviour of these *assemblées*（组合体）with an accuracy increasing as the number of atoms involved increases. It is in that way that the events acquire truly orderly features. All the physical and chemical laws that are known to play an important part in the life of organisms are of this statistical kind; any other kind of lawfulness and orderliness that one might think of is being perpetually disturbed and made inoperative by the unceasing heat motion of the atoms.

THEIR PRECISION IS BASED ON THE LARGE NUMBER OF ATOMS INTERVENING FIRST EXAMPLE (PARAMAGNETISM)

Let me try to illustrate this by a few examples, picked somewhat at random out of thousands, and possibly not just the best ones to appeal to a reader who is learning for the first time about this condition of things—a condition which in modern physics and chemis-

基于原子统计学的物理学定律只是近似的

为什么不能在这种情况下实现呢，由少量原子构成有机体，并且有机体能够敏感地对一个或有限几个原子的作用反应？

因为我们知道，所有的原子每时每刻都在进行着根本无序的热运动，也就是说，这与它们的有序行为是相反的，并且不允许在少量原子之间发生的事件遵循任何一条我们认可的定律。只有在巨大数量的原子共同参与的情况下，统计学定律才开始产生作用并控制这些组合体的系统行为，并且它的精确性随着参与原子的数目增加而提高。正是通过这种方式，才能形成真正有序的特征。现有已知的有机体生命活动中，所有起重要作用的物理学和化学的定律，都与这种统计学吻合，任何人们所能设想的其他类型的规律和秩序，总是会被原子不停歇的热运动所扰乱，变得难以协调。

物理定律的精确度需要巨量原子的参与　第一例：顺磁性

让我尝试用几个例子来阐明这一点，这是从数千个例子中随机挑选出来的，也许不一定是最好的例子，以吸引第一次接触这些条件的读者——这里所说的条件是在现代物理学和化学中最基本的概念，就像生物学中所说的生命体都是由细胞组成

try is as fundamental as, say, the fact that organisms are composed 11 of cells is in biology, or as Newton's Law in astronomy, or even as the series of integers, 1, 2, 3, 4, 5, ···in mathematics. An entire newcomer should not expect to obtain from the following few pages a full understanding and appreciation of the subject, which is associated with the illustrious names of Ludwig Boltzmann and Willard Gibbs and treated in textbooks under the name of 'statistical thermodynamics'.

If you fill an oblong quartz tube with oxygen gas and put it into a magnetic field, you find that the gas is magnetized.[1] The magnetization is due to the fact that the oxygen molecules are little magnets and tend to orientate themselves parallel to the field, like a compass needle. But you must not think that they actually all turn parallel. For if you double the field, you get double the magnetization in your oxygen body, and that proportionality goes on to extremely high field strengths, the magnetization increasing at the rate of the field you apply.

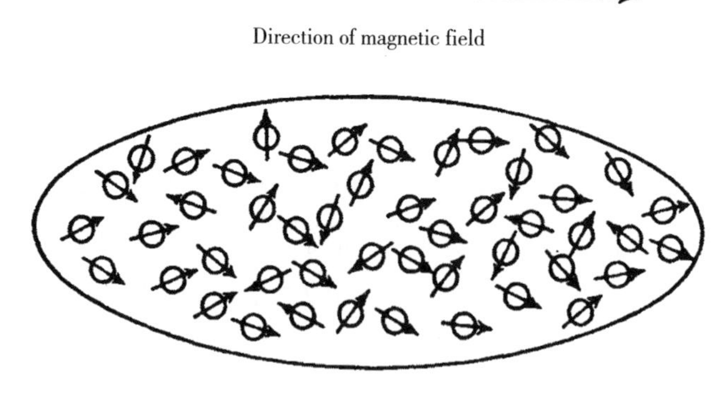

Fig. 1. Paramagnetism.

1. A gas is chosen, because it is simpler than a solid or a liquid; the fact that the magnetization is in this case extremely weak, will not impair the theoretical considerations.

的，天文学中的牛顿定律，甚至是数学中的整数序列1、2、3、4、5…等此类事实一样。一个刚刚涉足该领域的新手，不应该期望读了本书的下面几页就能完全领会这个主题。这个主题是与路德维希·玻尔兹曼和威拉德·吉布斯这两个著名的名字联系在一起的，在教科书中被冠以“统计热力学”之名。

如果你将一个椭圆形的石英管注满氧气，并把它放入一个磁场之中，你会发现气体被磁化了[1]。这是由于氧分子是一些小的磁体，倾向于使自己与磁场平行，就像一个罗盘针。但是你一定别认为它们全部与磁场平行了。因为如果你把磁场加倍，容器里的氧气磁化也会加倍，这种比例关系持续到极高的场强。磁化强度会随着你施加的磁场强度的增加速率而同步增长。

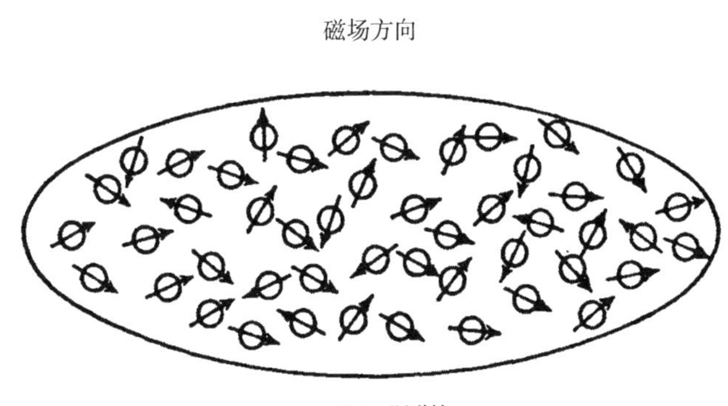

图 1　顺磁性

1. 之所以选用气体，是由于它比固体或液体简单一些，虽然事实上发生的磁化作用极弱，但不影响从理论上对这一现象的考虑。

This is a particularly clear example of a purely statistical law. The orientation the field tends to produce is continually counteracted by the heat motion, which works for random orientation. The effect of this striving is, actually, only a small preference for acute over obtuse angles between the dipole axes and the field. Though the single 12 atoms change their orientation incessantly, they produce on the average (owing to their enormous number) a constant small preponderance of orientation in the direction of the field and proportional to it.

这是纯统计学定律的一个特别明显的例子。磁场对磁化方向的影响不断被热运动抵消，后者倾向于随机取向。这一竞争机制作用的结果是，实际上，从偶极轴和场之间的夹角来看，锐角比钝角更具有那么一点倾向性。虽然单原子在不停地改变方向，但是平均来看（由于它们数量巨大），它们在场的方向上产生了一个恒定的微小优势，并与它呈正相关。

This ingenious explanation is due to the French physicist P. Langevin. It can be checked in the following way. If the observed weak magnetization is really the outcome of rival tendencies, namely, the magnetic field, which aims at combing all the molecules parallel, and the heat motion, which makes for random orientation, then it ought to be possible to increase the magnetization by weakening the heat motion, that is to say, by lowering the temperature, instead of reinforcing the field. That is confirmed by experiment, which gives the magnetization inversely proportional to the absolute temperature, in quantitative agreement with theory (Curie's law). Modern equipment even enables us, by lowering the temperature, to reduce the heat motion to such insignificance that the orientating tendency of the magnetic field can assert itself, if not completely, at least sufficiently to produce a substantial fraction of 'complete magnetization'. In this case we no longer expect that double the field strength will double the magnetization, but that the latter will increase less and less with increasing field, approaching what is called 'saturation'. This expectation too is quantitatively confirmed by experiment.

Notice that this behaviour entirely depends on the large numbers of molecules which co-operate in producing the observable magnetization. Otherwise, the latter would not be constant at all, but would, by fluctuating quite irregularly from one second to the next, bear witness to the vicissitudes of the contest between heat motion and field.

这个巧妙的解释归功于法国物理学家保罗·郎之万。这可以用下述的方法来验证：

如果观察到的弱磁化现象确实是两种对抗趋势平衡的结果，就是说磁场力图使所有分子与之平行，而热运动则使它们随机取向，那就应该有可能通过减弱热运动来增强磁化，也就是说，降低温度就能增加磁化强度，而不需要增强磁场。

实验已经证明了这一点，实验结果是磁化强度与绝对温度呈逆相关，在定量上与预期的理论（居里定律）一致。现代的实验装置甚至能让我们通过降低温度，把热运动降到极度微弱，以至充分显示出磁场的固定取向的趋势，即使不能全部，至少也是相当大一部分“完全磁化”。这种情况下，我们不再期望场强的加倍会使磁化加倍，而是希望看到随着场的进一步增强，磁化的增强愈来愈小，接近所谓的“饱和”。定量实验也证实了这一预期。

要注意的是，这种情况的出现依赖于大量的分子，这些分子的协同作用能产生可观察到的磁化现象。否则，磁化就不会是恒定的，而是每分每秒都在无休止地、不规则地变动着。这是热运动和磁场两者之间此消彼长、相互制衡的证明。

SECOND EXAMPLE (BROWNIAN MOVEMENT, DIFFUSION)

If you fill the lower part of a closed glass vessel with fog, consisting of minute droplets, you will find that the upper boundary of the fog gradually sinks, with a well-defined velocity, determined by 13 the viscosity of the air and the size and the specific gravity of the droplets. But if you look at one of the droplets under the microscope you find that it does not permanently sink with constant velocity, but performs a very irregular movement, the so-called Brownian movement, which corresponds to a regular sinking only on the average.

Now these droplets are not atoms, but they are sufficiently small and light to be not entirely insusceptible to the impact of one single molecule of those which hammer their surface in perpetual

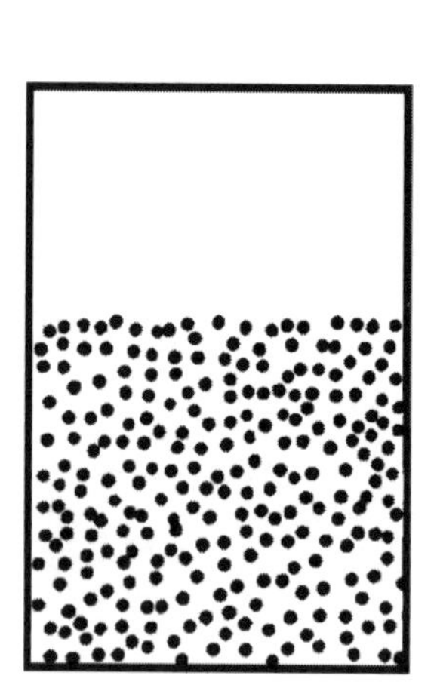

Fig. 2. Sinking fog.

Fig. 3. Brownian movement of a sinking droplet.

第二例：布朗运动，分子扩散

如果你把一个密闭玻璃容器的下部灌满微滴组成的雾状物，你将发现雾的上边界会按一定的速度逐渐下沉，其下沉速度取决于空气的黏度、微滴的大小和比重。但是，如果你在显微镜下观察任意一粒微滴，你就会发现它并不是一直以恒定的速度下沉，而是十分不规则地运动，即所谓的布朗运动。这种运动只是相对于平均水平的规律性下沉。

虽然这些微滴不是原子，但它们如此之小、如此之轻，对单个分子连续不断撞击其表面的效应并非完全没有反应。于是，它们就这样被撞来撞去，一般来说只能受重力的影响而下沉。

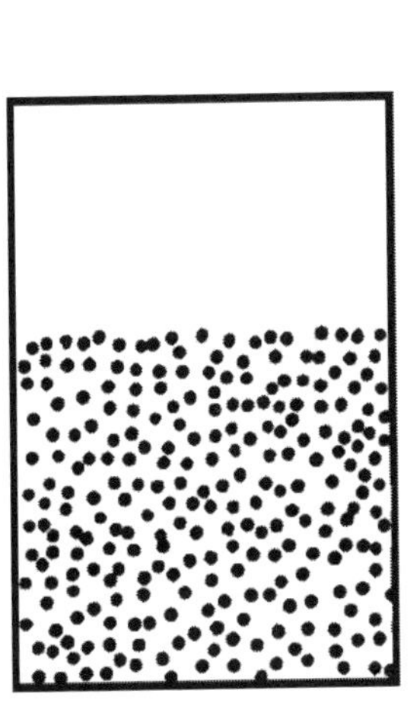

图2　下沉的雾滴

图3　某一下沉雾滴的布朗运动

impacts. They are thus knocked about and can only on the average follow the influence of gravity.

This example shows what funny and disorderly experience we should have if our senses were susceptible to the impact of a few molecules only. There are bacteria and other organisms so small that they are strongly affected by this phenomenon. Their movements are determined by the thermic whims of the surrounding medium; they have no choice. If they had some locomotion (移位) of their own they might nevertheless succeed in getting from one place to another—but with some difficulty, since the heat motion tosses them like a small boat in a rough sea.

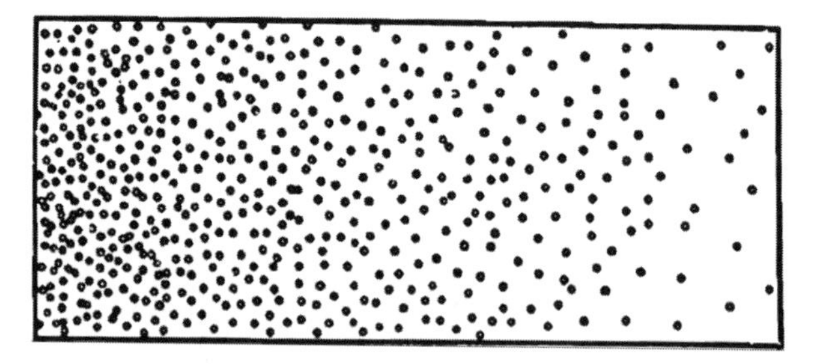

Fig. 4. Diffusion from left to right in a solution of varying concentration.

这个例子说明，假如人类的感官能感受到区区几个分子的碰撞，我们将会获得何等可笑而混乱的体验。像细菌及很多其他微小的生命体，它们都受到这种现象的显著影响。它们的运动受到周围介质中分子的热运动的影响，而自己并没有多少选择。如果它们能自主运动，它们可能会成功地从一个地方移动到另一个地方，但实际上这有点困难，因为热运动将它们抛来抛去，就像惊涛骇浪中的一叶扁舟。

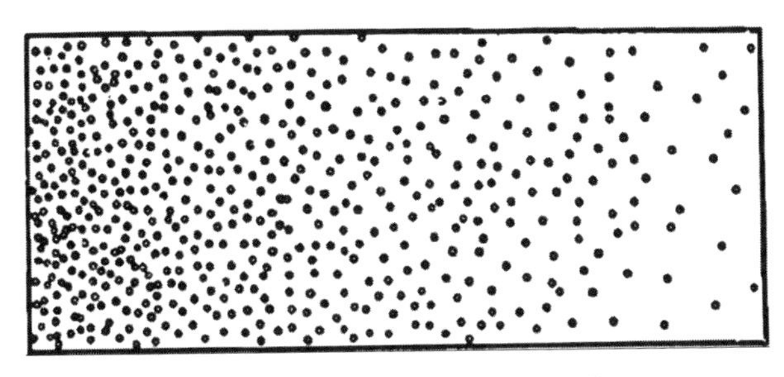

图4　溶液浓度不均匀时从左向右的扩散

A phenomenon very much akin to Brownian movement is that of *diffusion*. Imagine a vessel filled with a fluid, say water, with a small amount of some coloured substance dissolved in it, say potassium permanganate, not in uniform concentration, but rather as in Fig. 4, where the dots indicate the molecules of the dissolved substance (permanganate) and the concentration diminishes from left to right. If you leave this system alone a very slow process of 'diffusion' sets in, the permanganate spreading in the direction from left to right, that is, from the places of higher concentration towards the places of lower concentration, until it is equally distributed through the water.

The remarkable thing about this rather simple and apparently not particularly interesting process is that it is in no way due, as one might think, to any tendency or force driving the permanganate molecules away from the crowded region to the less crowded one, like the population of a country spreading to those parts where there is more elbow-room. Nothing of the sort happens with our permanganate molecules. Every one of them behaves quite independently of all the others, which it very seldom meets. Everyone of them, whether in a crowded region or in an empty one, suffers the same fate of being continually knocked about by the impacts of the water molecules and thereby gradually moving on in an unpredictable direction—sometimes towards the higher, sometimes towards the lower, concentrations, sometimes obliquely. The kind of motion it performs has often been compared with that of a blindfolded person

一种很像布朗运动的分子运动现象是扩散。请试想，在一个装满液体——就说水吧——的容器中，溶解了少量的有色物质，比如高锰酸钾，其浓度并不均匀，如图4所示，图中的小点代表溶质（高锰酸盐）分子，浓度由左向右逐步递减。如果你不去管它，高锰酸盐分子就会开始非常缓慢的"扩散"过程，向从左到右的方向扩散，即从高浓度处向低浓度处扩散，直到均匀地分布于水中。

这个那么简单明了，显然并不特别有趣的过程的非凡之处在于：它绝非人们想象的那样，高锰酸盐分子在某种趋势或是力量的驱使下，从密集区迁移到不太密集的区域，如同一个国家的人口向地广人稀的地区流动那样。对于我们的高锰酸盐分子来说，这种情况根本不会发生。每一个高锰酸盐分子对所有其他的高锰酸盐分子而言，都在完全独立地运动，很少和其他高锰酸盐分子相碰撞。可是，每一个高锰酸盐分子，无论是在密集区还是在空旷区，都会与水分子不断撞击，从而向着无法预估的方向逐渐移动——有时朝高浓度的方向，有时朝低浓度的方向，有时则是斜向移动。这个样子的运动经常被比喻为，一个被蒙着眼睛的人站在开阔的空地上，心里只有某种"行走"的欲望，可是并没有往任何特定方向的意向，因而他会不断地变换着他的路线。

on a large surface imbued with a certain desire of 'walking', but without any preference for any particular direction, and so changing his line continuously.

That this random walk of the permanganate molecules, the same for all of them, should yet produce a regular flow towards the smaller concentration and ultimately make for uniformity of distribution, is at first sight perplexing—but only at first sight. If you contemplate in Fig. 4 thin slices of approximately constant concentration, the permanganate molecules which in a given moment are contained in a particular slice will, by their random walk, it is true, be carried with equal probability to the right or to the left. But precisely in consequence of this, a plane separating two neighbouring slices will be crossed by more molecules coming from the left than in the opposite direction, simply because to the left there are more molecules engaged in random walk than there are to the right. And as long as that is so the balance will show up as a regular flow from left to right, until a uniform distribution is reached.

高锰酸盐分子的这种随机游走，对所有的高锰酸盐分子都是一样的，都会产生一种有规则的朝低浓度方向流动的趋势，最终均匀地分布在水中。乍看起来，这是令人困惑不解的，但仅仅是乍看起来而已。如果你把图4想象为一层层浓度几乎恒定的薄片，某一瞬间里，某一薄片所含的高锰酸盐分子，由于随机游走，确实有相同的几率被带到右边或左边。但正是由于这一点，通过某一分隔两层相邻薄片的平面的分子，来自左面的要多于来自右面的，这是由于左面比右面有更多的分子参与随机游走。只要是这种情况，均衡状态就将表现为一种自左到右的有规则的流动，直至达到均匀分布。

When these considerations are translated into mathematical language the exact law of diffusion is reached in the form of a partial differential equation

$$\frac{\partial \rho}{\partial t}=D\nabla^2\rho,$$

16 which I shall not trouble the reader by explaining, though its meaning in ordinary language is again simple enough.[1] The reason for mentioning the stern 'mathematically exact' law here, is to emphasize that its physical exactitude must nevertheless be challenged in every particular application. Being based on pure chance, its validity is only approximate. If it is, as a rule, a very good approximation, that is only due to the enormous number of molecules that co-operate in the phenomenon. The smaller their number, the larger the quite haphazard deviations we must expect—and they can be observed under favourable circumstances.

1. To wit: the concentration at any given point increases (or decreases) at a time rate proportional to the comparative surplus (or deficiency) of concentration in its infinitesimal environment. The law of heat conduction is, by the way, of exactly the same form, 'concentration' having to be replaced by 'temperature'.

当这些想法被翻译成数学语言时，扩散的确切规律就可以用偏微分方程的形式来表述：

$$\frac{\partial \rho}{\partial t}=D\nabla^2\rho。$$

我在这里就不作解释了，省得给读者带来理解这个方程的困难，尽管它也可以用足够通俗的语言来描述[1]。之所以在此提及严格的“在数学上精确的”定律，是为了强调它在物理学上的精确性必定会在每一个具体的应用中受到挑战。由于这一定律纯粹建立在概率的基础上，所以它只能说是近似有效的。照此，按常规来说，如果它是一个非常好的近似，那么也是由于这一现象中共同作用的原子数目非常庞大。我们必须预计到，原子数目越少，偶然偏差就越大，而且在合适的条件下，这些偏差是可以被观察到的。

1．也就是说：任何一处（点）的浓度都随着一定的时间速率而增加（或减小），这个时间速率与这一处（点）的周围无限小区域内的浓度的相对增加（或减少）是呈正相关的。顺便说一下，热传导定律的形式与此完全相同，只不过需要将“浓度”替换成“温度”。

THIRD EXAMPLE(LIMITS OF ACCURACY OF MEASURING)

The last example we shall give is closely akin to the second one, but has a particular interest. A light body, suspended by a long thin fibre in equilibrium orientation, is often used by physicists to measure weak forces which deflect it from that position of equilibrium, electric, magnetic or gravitational forces being applied so as to twist it around the vertical axis. (The light body must, of course, be chosen appropriately for the particular purpose.) The continued effort to improve the accuracy of this very commonly used device of a 'torsional balance', has encountered a curious limit, most interesting in itself. In choosing lighter and lighter bodies and thinner and longer fibres—to make the balance susceptible to weaker and weaker forces—the limit was reached when the suspended body became noticeably susceptible to the impacts of the heat motion of the surrounding molecules and began to perform an incessant, irregular 'dance' about its equilibrium position, much like the trembling of the droplet in the second example. Though this behaviour sets no absolute limit to the accuracy of measurements obtained with the balance, it sets a practical one. The uncontrollable effect of the heat 17 motion competes with the effect of the force to be measured and makes the single deflection observed insignificant. You have to multiply observations, in order to eliminate the effect of the Brownian movement of your instrument. This example is, I think, particularly illuminating in our present investigation. For our organs of sense, af-

第三例：精确测量的限度

我们要讨论的最后一个例子同第二个例子很相似，但它具有独特的意义。

一个很轻的小球，由一根长而细的纤丝悬吊在空中，使之趋向于平衡，经常被物理学家用来测量使小球偏离平衡位置的弱力，施加的电、磁或引力使其绕垂直轴扭曲（当然，必须根据特定的目的来选择这种轻小物体）。在持之以恒地改进这种常用的“扭力天平”装置的精确度时，却遇到了一个奇怪的限制，这本身是最有趣的。在选用愈来愈轻的物体和更细更长的纤丝时，这个平衡就能够被越来越弱的力所影响。当悬挂的物体能明显地感受到周围分子热运动的冲击时，就达到了它测量精度的极限，它就会在平衡位置附近不停地无规则地“乱舞”，很像第二个例子中的微滴颤动那样。虽然这样做并没有给天平测量的精确性带来任何绝对的限制，但它带来了实际限制。

热运动的不可控效应与被测量的力的效应相竞争，使观察到的单个偏差值不显著。这就使得我们需要多次观察，以消除布朗运动对实验仪器的影响。我认为，这个例子对我们目前的

ter all, are a kind of instrument. We can see how useless they would be if they became too sensitive.

THE $\sqrt{n}$ RULE

So much for examples, for the present. I will merely add that there is not one law of physics or chemistry, of those that are relevant within an organism or in its interactions with its environment, that I might not choose as an example. The detailed explanation might be more complicated, but the salient（关键的）point would always be the same and thus the description would become monotonous（单调的）.

But I should like to add one very important quantitative statement concerning the degree of inaccuracy to be expected in any physical law, the so-called $\sqrt{n}$ law. I will first illustrate it by a simple example and then generalize it.

研究是极具启发性的，毕竟我们的感觉器官也相当于一种仪器。如果感官变得过于灵敏，它反而没有什么用处了。

$\sqrt{n}$ 定律

暂且就举这么多例子吧。我只想再补充一点，凡是同生命体有关的，或者与生命体和环境相互作用有关的物理学或化学定律中，没有一条我不能选择作为例子。详细的解释也许更复杂些，但要点都一样，因此再说下去就会变得单调乏味了。

不过，我还是想补充一个非常重要的定量规律，它与所有物理学定律的不准确程度有关，即所谓的$\sqrt{n}$定律。我将首先通过一个简单的例子来说明它，然后再加以概括。

If I tell you that a certain gas under certain conditions of pressure and temperature has a certain density, and if I expressed this by saying that within a certain volume (of a size relevant for some experiment) there are under these conditions just n molecules of the gas, then you might be sure that if you could test my statement in a particular moment of time, you would find it inaccurate, the departure being of the order of $\sqrt{n}$. Hence if the number $n = 100$, you would find a departure of about 10, thus relative error = 10%. But if $n = 1$ million, you would be likely to find a departure of about 1,000, thus relative error = $\frac{1}{10}$%. Now, roughly speaking, this statistical law is quite general. The laws of physics and physical chemistry are inaccurate within a probable relative error of the order of $1/\sqrt{n}$, where n is the number of molecules that co-operate to bring about that law—to produce its validity within such regions of space or time (or both) that matter, for some considerations or for some particular experiment.

18 You see from this again that an organism must have a comparatively gross structure in order to enjoy the benefit of fairly accurate laws, both for its internal life and for its interplay with the external world. For otherwise the number of co-operating particles would be too small, the ‘law’ too inaccurate. The particularly exigent (苛求的) demand is the square root. For though a million is a reasonably large number, an accuracy of just 1 in 1,000 is not overwhelmingly good, if a thing claims the dignity of being a ‘Law of Nature’.

如果我告诉你，某种气体在一定压力和温度条件下具有一定的密度，或者换种说法，在一定的体积内（适合一些实验的需要），在一定的压力和温度条件下，正好有n个气体分子，那么你就可以肯定，若你要在某一特定时刻对我这个说法进行验证，你将会发现这个说法是不准确的，分子数量的偏差将是$\sqrt{n}$的量级。因此，如果数量$n=100$，你将发现偏差大约是10，相对误差为10%。可是，如果$n=1,000,000$，你很可能会发现偏差约为1,000，因而相对误差为0.1%。现在，粗略地说，这一统计学定律是普遍成立的。物理学和物理化学定律的不准确程度总是介于 $1/\sqrt{n}$ 量级的相对误差范围之内，其中n是为使定律成立而设计的参与合作的分子的个数，对于某些假设或特定的实验来说，在一定的空间或时间（或同一空间和时间）范围内，该定律是适用的。

由此你又可以看出，一个生命体必须拥有一个相对巨大的结构，才能从相对精确的规律中获益，无论是它的内部，还是它与外部世界的相互作用。否则，参与合作的粒子数目太小的话，“定律”就不会那么精确了。需要特别注意的是这里的平方根。因为，尽管100万是一个相当大的数字，可是其精确度只有千分之一，这样的精确度对于一条所谓的“自然定律”来说还是远远不够的。

CHAPTER 2

The Hereditary Mechanism

Das Sein ist ewig; denn Gesetze
Bewahren die lebend'gen Schätze,
Aus welchen sich das All geschmückt.[1]

GOETHE

THE CLASSICAL PHYSICIST'S EXPECTATION, FAR FROM BEING TRIVIAL, IS WRONG

19 Thus we have come to the conclusion that an organism and all the biologically relevant processes that it experiences must have an extremely 'many-atomic' structure and must be safeguarded against haphazard, 'single-atomic' events attaining too great importance. That, the 'naïve physicist' tells us, is essential, so that the organism may, so to speak, have sufficiently accurate physical laws on which to draw for setting up its marvellously regular and well-ordered working. How do these conclusions, reached, biologically speaking,

1. Being is eternal; for laws there are to conserve the treasures of life on which the Universe draws for beauty.

第二章
遗传机制

存在跨越了岁月之长河
定律保存了万物之精华，
宇宙汲取了生命之美妙。
——歌德

肤浅与谬误——经典物理学家的期待

因此，我们可以得出这样的结论，一个生命体和它所有生物学相关过程，都必须具备足够的“多原子”结构，以防止偶发的“单原子”事件产生太大的影响。“天真的物理学家”告诉我们，这一点是至关重要的，因此才可以说，生命体可能有足够精确的物理学定律，可以据此建立其极具规律和秩序的功能。这些结论，从生物学角度来看，是先验的（即，从纯粹的物理学观点得出的），是否和实实在在的生物学事实相符呢？

a priori (that is, from the purely physical point of view), fit in with actual biological facts?

At first sight one is inclined to think that the conclusions are little more than trivial. A biologist of, say, thirty years ago might have said that, although it was quite suitable for a popular lecturer to emphasize the importance, in the organism as elsewhere, of statistical physics, the point was, in fact, rather a familiar truism. For, naturally, not only the body of an adult individual of any higher species, but every single cell composing it contains a 'cosmical' number of single atoms of every kind. And every particular physiological process that we observe, either within the cell or in its interaction with the environment, appears—or appeared thirty years ago—to involve such enormous numbers of single atoms and single atomic processes that all the relevant laws of physics and physical chemistry would be safeguarded even under the very exacting demands of statistical physics in respect of 'large numbers'; this demand I illustrated just now by the $\sqrt{n}$ rule.

乍看起来，人们往往认为这些结论是无关紧要的。比如说，一位30年前的生物学家可能会说，尽管这对于一个受大众欢迎的演说家来说，强调统计物理学在生物学和其他领域中的重要性是非常适合的，但是实际上，这种观点是“老生常谈”罢了。因为，自然而然地，不仅仅是任何高等物种的成年个体的身体，而且构成其躯体的每一个细胞都包含着“天文数字”的各色各样的单原子。而我们观察的每一个特定的生理过程，无论是在细胞内部还是在细胞与环境的相互作用中，似乎——或30年前就这样——同样涉及数目庞大的单原子和单原子过程，以至于所有相关的物理学和物理化学定律，即使在统计物理学对巨大数字的严格要求下，依然适用于几乎所有；我刚才用$\sqrt{n}$定律就说明了这一点。

Today, we know that this opinion would have been a mistake. As we shall presently see, incredibly small groups of atoms, much too small to display exact statistical laws, do play a dominating role in the very orderly and lawful events within a living organism. They have control of the observable large-scale features which the organism acquires in the course of its development, they determine important characteristics of its functioning; and in all this very sharp and very strict biological laws are displayed.

I must begin with giving a brief summary of the situation in biology, more especially in genetics—in other words, I have to summarize the present state of knowledge in a subject of which I am not a master. This cannot be helped and I apologize, particularly to any biologist, for the dilettante（外行的）character of my summary. On the other hand, I beg leave to put the prevailing ideas before you more or less dogmatically（教条主义地）. A poor theoretical physicist could not be expected to produce anything like a competent survey of the experimental evidence, which consists of a large number of long and beautifully interwoven series of breeding experiments of truly unprecedented ingenuity on the one hand and of direct observations of the living cell, conducted with all the refinement of modern microscopy, on the other.

今天，我们知道，这种观点可能是一个错误。正如我们现在将会看到的，那些小得令人难以置信的原子团，小到无法显示精确的统计学定律，但它们却在活机体内非常有序的、符合规律的活动中发挥着主导作用。它们控制着生命体在发育过程中可观察到的宏观特征，并且决定着生命体功能的重要特征；在所有这些过程中，都展示了非常明确而又严格的生物学规律。

我必须首先简要地概括一下生物学的发展现状，特别是从遗传学的角度出发——换言之，我必须要对目前的知识现状作一概括，尽管这并不是我精通的学科。这是不得已而为之，我要为我说的概括中的一些外行话道歉，特别是向所有的生物学家道歉；另一方面，也请允许我多少带点教条地向大家介绍一下普遍的观点。是不指望一个蹩脚的理论物理学家能写出一份合格的实验数据调查报告的。这份调查报告一方面包括大量日积月累的实验数据与具有真正前所未有的独创性的育种实验结果，另一方面包括用最精密的现代显微镜技术对活细胞的直接观察所得出的结论。

THE HEREDITARY CODE-SCRIPT (CHROMOSOMES)

Let me use the word 'pattern' of an organism in the sense in which the biologist calls it 'the four-dimensional pattern', meaning not only the structure and functioning of that organism in the adult, 21 or in any other particular stage, but the whole of its ontogenetic（个体发育的）development from the fertilized egg cell to the stage of maturity, when the organism begins to reproduce itself. Now, this whole four-dimensional pattern is known to be determined by the structure of that one cell, the fertilized egg. Moreover, we know that it is essentially determined by the structure of only a small part of that cell, its nucleus. This nucleus, in the ordinary 'resting state' of the cell, usually appears as a network of chromatine,[1] distributed over the cell. But in the vitally important processes of cell division (mitosis and meiosis, see below) it is seen to consist of a set of particles, usually fibre-shaped or rod-like, called the chromosomes, which number 8 or 12 or, in man, 48. But I ought really to have written these illustrative numbers as 2 × 4, 2 × 6, ···, 2 × 24, ···, and I ought to have spoken of two sets, in order to use the expression in the customary meaning of the biologist. For though the single chromosomes are sometimes clearly distinguished and individualized by shape and size, the two sets are almost entirely alike. As we have shall see in a moment, one set comes from the mother (egg cell), one

1. The word means 'the substance which takes on colour', viz. in a certain dyeing process used in microscopic technique.

染色体：遗传的密码本

请允许我使用生命体的“模式”一词，生物学家称之为“四维模式”，它不仅意味着生命体在成年之后或其他特定阶段的结构和功能，而且表示从受精卵细胞到成熟阶段的整个个体发育过程，即生物体开始自我繁殖的阶段。

现在，我们已经知道整个四维模式是由一个单细胞，即受精卵的结构决定的。而且，我们知道，本质上是由受精卵的很小一部分结构，即它的细胞核所决定的。这个细胞核在体细胞的正常“休眠期”内，其内的染色质[1]通常呈网状分布在细胞中。但在至关重要的细胞分裂（有丝分裂和减数分裂，见下文）过程中，可以观察到染色质是由一组微粒组成的，通常呈纤维状或棒状结构，被称为染色体。其数量有8条或12条，人类则有48条[2]。

不过，我应该把这些染色体数目写成2 × 4，2 × 6，……，2 × 24，……，称它们为两组染色体，以按照生物学家的习惯使用这一表述。尽管单个染色体有时会因它的形状和大小而被明确区分和个性化，但这两组染色体几乎是一模一样的。我们很快就能看到，这是由于它们之中的一组来自母本（卵细胞），

1. 这个名词的含义是指在显微技术的某个染色步骤中“能染上颜色的物质”。
2. 此处原文为48条，1956年华裔遗传学家蒋有兴（1919—2001）与瑞典遗传学家阿尔伯特·莱万（1905—1997）已证明人类有46条染色体。

from the father (fertilizing spermatozoon). It is these chromosomes, or probably only an axial skeleton fibre of what we actually see under the microscope as the chromosome, that contain in some kind of code-script the entire pattern of the individual's future development and of its functioning in the mature state. Every complete set of chromosomes contains the full code; so there are, as a rule, two copies of the latter in the fertilized egg cell, which forms the earliest stage of the future individual.

In calling the structure of the chromosome fibres a code-script we mean that the all-penetrating mind, once conceived by Laplace, to which every causal connection lay immediately open, could tell from their structure whether the egg would develop, under suitable conditions, into a black cock or into a speckled（布满小斑点的）hen, into a fly or a maize plant, a rhododendron（杜鹃花）, a beetle, a mouse or a woman. To which we may add, that the appearances of 22 the egg cells are very often remarkably similar; and even when they are not, as in the case of the comparatively gigantic eggs of birds and reptiles, the difference is not so much in the relevant structures as in the nutritive material which in these cases is added for obvious reasons.

But the term code-script is, of course, too narrow. The chromosome structures are at the same time instrumental in bringing about the development they foreshadow. They are law-code and executive power—or, to use another simile（明喻）, they are architect's plan and builder's craft—in one.

另一组来自父本（授精的精子）。正是这些染色体，或者说可能只是我们在显微镜下看到的染色体的一种骨架纤丝，它们以某种密码本的形式包含了个体未来发育的全部模式和成熟状态下的功能。每套完整的染色体都包含完整的密码，因此按常规，在受精的卵细胞里通常含有两份密码本，从而形成未来个体发育的早期阶段。

把染色体纤丝结构称为密码本，我们在这里是指拉普拉斯曾经设想过的能洞察未来的意念，它对每一种因果联系都能立即阐明，它可以从卵的结构中看出，在适宜的条件下，这个卵将发育成一只黑公鸡还是一只芦花母鸡，是长成一只苍蝇还是一棵玉米植株，是一株杜鹃花还是一只甲虫，是一只老鼠又或是一位女士。除此之外，我们还可以补充的一点就是卵细胞的外观往往非常相似；即使外观不相似，比如鸟类和爬行类的卵相对来说就大得多，其区别也不是在于其相关结构不同，而在于其营养成分的差别，因为在这些卵中，营养成分的差异是显而易见的。

当然，“密码本”一词太过于简单了。因为染色体结构同时也负责引导卵细胞按照它们的指令发育。也就是说，染色体是法律条文与执行能力的统一，打个比方，它集设计师的蓝图与建筑者的技艺于一身。

GROWTH OF THE BODY BY CELL DIVISION (MITOSIS)

How do the chromosomes behave in ontogenesis?[1]

The growth of an organism is effected by consecutive cell divisions. Such a cell division is called mitosis. It is, in the life of a cell, not such a very frequent event as one might expect, considering the enormous number of cells of which our body is composed. In the beginning the growth is rapid. The egg divides into two 'daughter cells' which, at the next step, will produce a generation of four, then of 8, 16, 32, 64, ···, etc. The frequency of division will not remain exactly the same in all parts of the growing body, and that will break the regularity of these numbers. But from their rapid increase we infer by an easy computation that on the average as few as 50 or 60 successive divisions suffice to produce the number of cells in a grown man—or, say, ten times the number,[2] taking into account the exchange of cells during lifetime. Thus, a body cell of mine is, on the average, only the 50th or 60th 'descendant' of the egg that was I.

1. Ontogenesis is the development of the individual, during its lifetime, as opposed to phylogenesis, the development of species within geological periods.
2. Very roughly, a hundred or a thousand (English) billions.

有丝分裂：细胞分裂与个体生长

在个体发育[1]过程中，染色体的行为又是如何？

一个有机体的生长是细胞连续分裂的结果。这样的细胞分裂称为有丝分裂。考虑到我们的身体是由大量的细胞组成的，在一个细胞的生命里，有丝分裂的次数并不像人们所想的那么多。组成我们个体的细胞数目巨大，在开始生长初期，受精卵生长迅速，并分裂成2个“子细胞”，下一步分裂成4个细胞，然后为8，16，32，64……但是，正在发育的身体各个部分中，细胞分裂频率并非保持完全一致，那样便会打破各个部分细胞数目的规律。不过，从细胞的迅速增长，我们通过简单的计算便可推断出，平均只需要50次或者60次的连续分裂，便足以产生出相当于一个成人的细胞数，或者说，将人一生中细胞的新陈代谢也考虑在内，大概是这个数目的10倍[2]。因此，平均来说，我现在身体内的每一个细胞都只是我那个最初的卵细胞的第50代或60代。

1. 个体发育是指一个个体在其生命周期中的发育；与之对应的是系统发育，指一个物种在不同地质年代内的发展。
2. 约为10^{11}个或10^{12}个。

IN MITOSIS EVERY CHROMOSOME IS DUPLICATED

How do the chromosomes behave on mitosis?

23 They duplicate—both sets, both copies of the code, duplicate. The process has been intensively studied under the microscope and is of paramount interest, but much too involved to describe here in detail. The salient point is that each of the two 'daughter cells' gets a dowry（嫁妆）of two further complete sets of chromosomes exactly similar to those of the parent cell. So all the body cells are exactly alike as regards their chromosome treasure.[1]

However little we understand the device we cannot but think that it must be in some way very relevant to the functioning of the organism, that every single cell, even a less important one, should be in possession of a complete (double) copy of the code-script. Some time ago we were told in the newspapers that in his African campaign General Montgomery made a point of having every single soldier of his army meticulously（一丝不苟地）informed of all his designs. If that is true (as it conceivably might be, considering the high intelligence and reliability of his troops) it provides an excellent analogy to our case, in which the corresponding fact certainly is literally true. The most surprising fact is the doubleness of the chromosome set, maintained throughout the mitotic divisions. That it is the outstanding feature of the genetic mechanism is most strikingly revealed by the one and only departure from the rule, which we have now to discuss.

---

1. The biologist will forgive me for disregarding in this brief summary the exceptional case of mosaics.

有丝分裂中染色体的加倍

在有丝分裂时，染色体行为如何？

它们会进行复制 —— 两组染色体和两套密码本都被复制了。这个过程在显微镜下已做了详尽的研究，极其有趣但又非常复杂，这里就不做详细的阐述了。这个过程的关键点是：两个“子细胞”中的每一个都得到了亲本细胞的“嫁妆”—— 与亲本几乎相同的全套染色体。因而，就人们的染色体的宝藏来说，所有的体细胞的染色体都是完全相同的[1]。

尽管对这种机制知之甚少，但我们可以肯定的是，它一定是通过某种途径同生命体的功能密切联系，因为每个单细胞，甚至那些不那么紧要的细胞，都具有密码本的全套（双倍的）拷贝。不久前，我在报纸上看到蒙哥马利将军在非洲战役中，要求他麾下的每一名战士都要了解他的极为细致的作战计划。如果那是真的（考虑到他的部队聪慧可靠，这个报道很有可能是真实的），那正好为我们的案例提供了一个完美的类比，在这个类比中，一个士兵就相当于一个细胞。最令人惊异的事实是，在整个有丝分裂过程中染色体始终保持双倍。这是遗传机制的突出特征，但是，细胞分裂也会有例外，我们接下来就将对这种独一无二的例外进行讨论。

1. 请生物学家原谅我在这个简短的归纳中没有提到嵌合体这样的例外。

REDUCTIVE DIVISION (MEIOSIS) AND FERTILIZATION (SYNGAMY)

Very soon after the development of the individual has set in, a group of cells is reserved for producing at a later stage the so-called gametes, the sperm cells or egg cells, as the case may be, needed for the reproduction of the individual in maturity. 'Reserved' means that they do not serve other purposes in the meantime and suffer many fewer mitotic divisions. The exceptional or reductive division (called meiosis) is the one by which eventually, on maturity, the gametes 24 are produced from these reserved cells, as a rule only a short time before syngamy is to take place. In meiosis the double chromosome set of the parent cell simply separates into two single sets, one of which goes to each of the two daughter cells, the gametes. In other words, the mitotic doubling of the number of chromosomes does not take place in meiosis, the number remains constant and thus every gamete receives only half—that is, only one complete copy of the code, not two, e.g. in man only 24, not 2 × 24 = 48.

减数分裂和受精

个体开始发育后不久，有一组细胞保留了下来，以便在发育的后续阶段能产生所谓的配子，视情况而定生成精子细胞或卵细胞，用于个体成熟后的生殖。“保留”是指它们在这段时期不用于其他目的，只是进行很少几次有丝分裂。作为例外或减少性的分裂（称为减数分裂），是在生命体的最终性成熟时从这些保留的细胞中产生配子的过程，通常只在交配发生前很短时间内进行。

在减数分裂中，亲本细胞双染色体组直接分成两个单染色体组，每一组染色体进入两个子细胞（即配子）中的一个。换句话说，在减数分裂中，并不像有丝分裂那样发生染色体数目的加倍，染色体数目是保持不变的，因此每个配子得到的只有一半——即一个完整的密码本拷贝，而不是两个，例如，人的配子中只有24条染色体，而不是2 × 24=48条染色体。

Cells with only one chromosome set are called haploid (from Greek ἁπλοῦς, single). Thus the gametes are haploid, the ordinary body cells diploid (from Greek διπλοῦς, double). Individuals with three, four, ···or generally speaking with many chromosome sets in all their body cells occur occasionally; the latter are then called triploid, tetraploid, ···, polyploid.

In the act of syngamy the male gamete (spermatozoon) and the female gamete (egg), both haploid cells, coalesce（结合）to form the fertilized egg cell, which is thus diploid. One of its chromosome sets comes from the mother, one from the father.

只有一组染色体的细胞叫作单倍体（来自希腊文ἁπλοῦς，意思是单一的）。因此，配子为单倍体的，正常的体细胞是二倍体（来自希腊文διπλοῦς，意思为双倍的）。偶尔出现的拥有三组、四组或多组染色体的体细胞的个体就称为三倍体、四倍体、……或多倍体。

在受精过程中，雄性配子（精子）和雌性配子（卵子）都是单倍体细胞，因此结合后形成的受精卵是二倍体的，一组染色体来自母本，一组来自父本。

HAPLOID INDIVIDUALS

25 One other point needs rectification. Though not indispensable for our purpose it is of real interest, since it shows that actually a fairly complete code-script of the 'pattern' is contained in every single set of chromosomes.

There are instances of meiosis not being followed shortly after by fertilization, the haploid cell (the 'gamete') undergoing meanwhile numerous mitotic cell divisions, which result in building up a complete haploid individual. This is the case in the male bee, the drone, which is produced parthenogenetically, that is, from non-fertilized and therefore haploid eggs of the queen. The drone has no

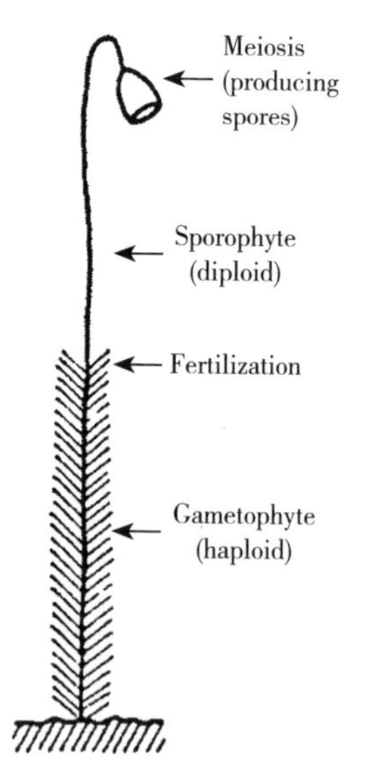

Fig. 5. Alternation of Generations.

单倍体个体

另一点需要澄清的是：尽管我们的研究目的不是单倍体，但它是真正有趣的，因为它表明，每一组染色体实际上都含有一个相当完整的“模式”密码本。

还有一些例子表明，减数分裂后并不立刻受精，其单倍体细胞（“配子”）也可以通过多次有丝分裂，从而形成一个完整的单倍体个体。雄蜂就是这样，它是孤雌生殖的，也就是说，它是由蜂后未受精的卵子——也就是单倍体的卵发育而来的。雄蜂

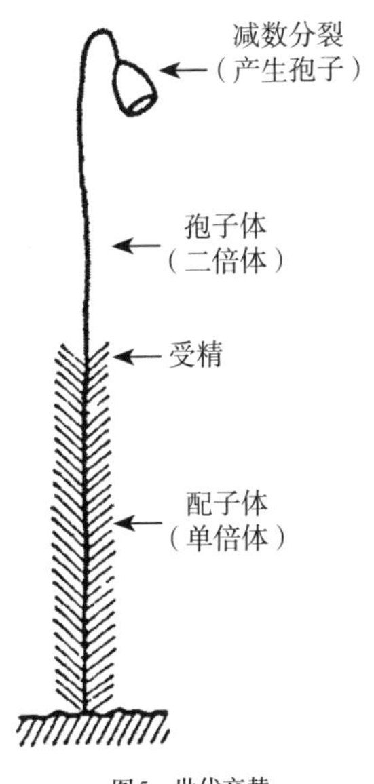

图5　世代交替

father! All its body cells are haploid. If you please, you may call it a grossly exaggerated spermatozoon; and actually, as everybody knows, to function as such happens to be its one and only task in life. However, that is perhaps a ludicrous point of view. For the case is not quite unique. There are families of plants in which the haploid gamete which is produced by meiosis and is called a spore in such cases falls to the ground and, like a seed, develops into a true haploid plant comparable in size with the diploid. Fig. 5 is a rough sketch of a moss, well known in our forests. The leafy lower part is the haploid plant, called the gametophyte, because at its upper end it develops sex organs and gametes, which by mutual fertilization produce in the ordinary way the diploid plant, the bare stem with the capsule at the top. This is called the sporophyte, because it produces, by meiosis, the spores in the capsule at the top. When the capsule opens, the spores fall to the ground and develop into a leafy stem, etc. The course of events is appropriately called alternation of generations. You may, if you choose, look upon the ordinary case, man and the animals, in the same way. But the 'gametophyte' is then as a rule a very short-lived, unicellular generation, spermatozoon or egg cell as the case may be. Our body corresponds to the sporophyte. Our 'spores' are the reserved cells from which, by meiosis, the unicellular generation springs.

是没有父亲的！它所有的体细胞都是单倍体。如果你愿意的话，可以把它看作是扩增了无数次的精子。事实上，众所周知，这样的功能恰好就是雄蜂一生中唯一的使命。

然而，这也许是一个可笑的观点。因为这种现象并不那么独特。有些科的植物，通过减数分裂产生单倍体配子，在这种情况下被称为孢子。孢子落在地上，就像一粒种子一样，可以发育成真正的单倍体植物，大小可与二倍体植物不相上下。图5就是一种苔藓类植物的草图，在森林中很常见。长有叶片的底部是单倍体植株，叫配子体，它的顶端可发育成性器官并产生配子，配子通过相互受精的方式，产生了二倍体植物，即裸露无叶的茎和顶部的孢子囊，称为孢子体，可通过减数分裂产生孢子。当孢子囊张开时，孢子落地，并发育成长为叶状茎。这个过程被恰当地称为世代交替。只要你愿意，你也可以用同样的方式来看待常见的情况，即人和动物也是如此。不过，“配子体”一般而言是寿命极短的单细胞一代，至于这种“配子体”是精子还是卵子要视情况而定。如果我们的身体相当于孢子体，那“孢子”就是那些被保留的细胞，通过减数分裂就可产生单细胞的配子。

THE OUTSTANDING RELEVANCE OF THE REDUCTIVE DIVISION

26 The important, the really fateful event in the process of reproduction of the individual is not fertilization but meiosis. One set of chromosomes is from the father, one from the mother. Neither chance nor destiny can interfere with that. Every man[1] owes just half of his inheritance to his mother, half of it to his father. That one or the other strain seems often to prevail is due to other reasons which we shall come to later. (Sex itself is, of course, the simplest instance of such prevalence.).

But when you trace the origin of your inheritance back to your grandparents, the case is different. Let me fix attention on my paternal set of chromosomes, in particular on one of them, say No.5. It is a faithful replica either of the No.5 my father received from his father or of the No.5 he had received from his mother. The issue was decided by a 50 : 50 chance in the meiosis taking place in my father's body in November 1886 and producing the spermatozoon which a few days later was to be effective in begetting me. Exactly the same story could be repeated about chromosomes Nos. 1, 2, 3, ⋯, 24 of my paternal set, and *mutatis mutandis* about every one of my maternal chromosomes. Moreover, all the 48 issues are entirely independent. Even if it were known that my paternal it chromosome No.5 came from my grandfather Josef Schrödinger, the No.7 still stands an equal chance of being either also from him, or from his wife Marie, née Bogner.

---

1. At any rate, every *woman*. To avoid prolixity, I have excluded from this summary the highly interesting sphere of sex determination and sex-linked properties (as, for example, so-called colour blindness).

减数分裂的卓越贡献

在个体繁殖过程中，重要的、真正命运攸关的事件并不是受精，而是减数分裂。

体细胞中的一组染色体来自父本，另一组来自母本，无论是偶然机遇还是命运决定，这都是无法干预的。每一个人[1]的染色体正好有一半遗传自他的母亲，另一半则遗传自他的父亲。至于父母哪一方占优势，那是由其他一些因素决定的，我们将在后面讨论（当然，性别本身也就是这种优势的最显而易见的例子）。

可是，当你将遗传的起源追溯到你的祖父母时，情况就不同了。现在，让我集中讨论父系的那一组染色体，特别是其中的一条，比如说，5号染色体。这条染色体是精确的复制品，要么是我父亲从他的父亲那里得到的，要么是从他的母亲那里得到的。在减数分裂中，这一事件发生的概率是50∶50，也就是说，1886年11月，我父亲体内发生了减数分裂并产生了精子，几天后，精子就有效地孕育了我。当然，我父母的染色体组中的1号、2号、3号、……、24号染色体都会重复同样的过程，并且我母系的每一条染色体也同样如此。而且，所有48条染色体都是各自完全独立的。即使我知道我父系的5号染色体来自我的祖父约瑟夫·薛定谔，而7号染色体仍然有相同的概率来自祖父，或者是来自祖母玛丽·博格纳。

1. 无论如何，每个女士也都是如此。为了避免冗长，我在这结论中省略了那些非常有趣的话题，如性别决定以及伴性性状（例如通常所说的色盲）。

CROSSING-OVER. LOCATION OF PROPERTIES

But pure chance has been given even a wider range in mixing the grandparental inheritance in the offspring than would appear 27 from the preceding description, in which it has been tacitly (默认地) assumed, or even explicitly stated, that a particular chromosome as a whole was either from the grandfather or from the grandmother; in other words that the single chromosomes are passed on undivided. In actual fact they are not, or not always. Before being separated in the reductive division, say the one in the father's body, any two 'homologous' chromosomes come into close contact with each other, during which they sometimes exchange entire portions in the way illustrated in Fig. 6. By this process, called 'crossing-over', two properties situated in the respective parts of that chromosome will be separated in the grandchild, who will follow the grandfather in one of them, the grandmother in the other one. The act of crossing-over, being neither very rare nor very frequent, has provided us with

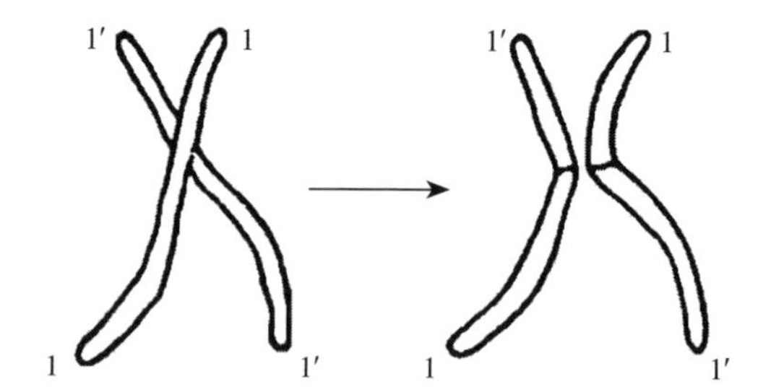

Fig. 6. *Crossing-over.* Left: the two homologous chromosomes in contact. Right: after exchange and separation.

染色体交换：性状的定位

但是，祖父祖母的混合遗传这一纯粹的概率事件在下一代出现的范围要比我们在前面描述的情况的范围大得多。如上所述，我们已经默认了，甚至明确地说，某条特定的整条染色体要么来自祖父，要么来自祖母；换句话说，这一条染色体是整条传递下去，而并不分割开来。然而事实并非如此，或者说并不总是这样。在减数分裂中，在染色体分离以前，比如说在父亲体内两条“同源”染色体都彼此紧靠在一起，在这段时间里，它们可能有机会进行整个片段的交换，如图6所示。

通过这种被称为“交换”的过程，分别位于那两条染色体上相应部位上的两个性状，就会在孙子孙女那一代分离，孙子孙女中就会一个性状像祖父，另一个性状像祖母。这种既非十分

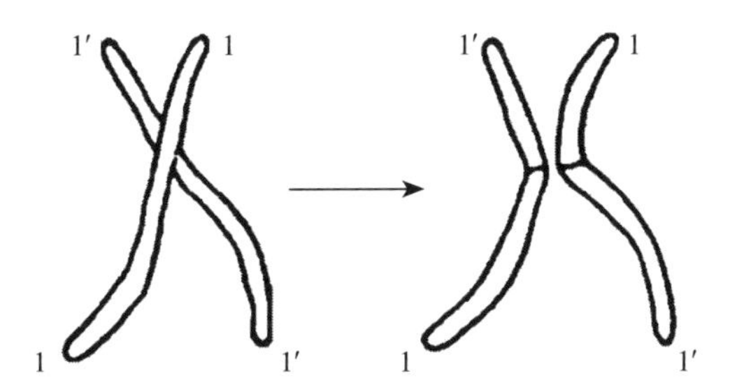

图6　染色体交换
左：彼此靠近的两条同源染色体　右：交换与分离之后

invaluable information regarding the location of properties in the chromosomes. For a full account we should have to draw on conceptions not introduced before the next chapter (e.g. heterozygosy, dominance, etc.); but as that would take us beyond the range of this little book, let me indicate the salient point right away.

If there were no crossing-over, two properties for which the same chromosome is responsible would always be passed on together, no descendant receiving one of them without receiving the other as well; but two properties, due to different chromosomes, would either stand a 50：50 chance of being separated or they would invariably be separated—the latter when they were situated in homologous chromosomes of the same ancestor, which could never go together.

罕见也非十分常见的交换过程，为我们研究染色体上不同性状的位置提供了十分宝贵的信息。若要作详细论述，我们就需要在讲下一章之前引入许多没有介绍过的概念（如杂合子、显性等）；但是，那么一来，就超过了我们这一本小书的范围，请让我在这里择要阐述。

假如没有交换，同一条染色体所决定的两个性状将恒定地一起遗传给下一代，后代不会只继承其中一个性状，而不继承另一个；可是，由不同的染色体决定的两个性状，要么以50∶50的概率被分开，要么永远被分开。当两个性状位于同一祖先的同源染色体上的时候，其后代就会出现性状分离，因为同源染色体上的两个性状是永远不会一起遗传给下一代的。

28 These rules and chances are interfered with by crossing-over. Hence the probability of this event can be ascertained by registering carefully the percentage composition of the off-spring in extended breeding experiments, suitably laid out for the purpose. In analysing the statistics, one accepts the suggestive working hypothesis that the 'linkage' between two properties situated in the same chromosome, is the less frequently broken by crossing-over, the nearer they lie to each other. For then there is less chance of the point of exchange lying between them, whereas properties located near the opposite ends of the chromosomes are separated by every crossing-over. (Much the same applies to the recombination of properties located in homologous chromosomes of the same ancestor.) In this way one may expect to get from the 'statistics of linkage' a sort of 'map of properties' within every chromosome.

These anticipations have been fully confirmed. In the cases to which tests have been thoroughly applied (mainly, but not only, *Drosophila*) the tested properties actually divide into as many separate groups, with no linkage from group to group, as there are different chromosomes (four in *Drosophila*). Within every group a linear map of properties can be drawn up which accounts quantitatively for the degree of linkage between any two out of that group, so that there is little doubt that they actually are located, and located along a line, as the rod-like shape of the chromosome suggests.

染色体交换干扰了这种遗传的规律和概率。因此，可以通过精心设计的大量的育种实验，仔细地统计和分析后代性状的百分比组成，并以此来确定交换的概率。在分析这些统计数据时，我们可以接受这样一个工作假设：位于同一条染色体上的两个性状（决定特性的两个基因）彼此靠得愈近，两者之间的"连锁"就越不容易被交换打乱。因为这样一来，在它们之间形成交换点的概率小了，而靠近染色体两端的性状则因为每一次交换而分离（这个道理，同样也适用于位于同一祖先的同源染色体上的性状重组）。通过这种方法，人们可以根据"连锁的统计"，绘制出每一条染色体所谓的"性状图"。

这些设想已经在实验中被完全证实。这些实验已经被广泛应用（主要用于果蝇，但不限于果蝇），被验证的性状实际上分成了不同的、相互独立的几个群，群与群之间没有连锁，因为这些群代表了不同的染色体（果蝇有4条染色体）。而在每个群的内部都可以绘制出性状分布的线性图谱，此图可以定量地描绘该群内任何两个性状之间的连锁程度，因此，毫无疑问，它们性状的位置被确定了，而且是线性排列，正如那个棒状染色体所表示的那样。

Of course, the scheme of the hereditary mechanism, as drawn up here, is still rather empty and colourless, even slightly naïve. For we have not said what exactly we understand by a property. It seems neither adequate nor possible to dissect into discrete 'properties' the pattern of an organism which is essentially a unity, a 'whole'. Now, what we actually state in any particular case is, that a pair of ancestors were different in a certain well-defined respect (say, one had blue eyes, the other brown), and that the offspring follows in this respect either one or the other. What we locate in the chromosome is the seat of this difference. (We call it, in technical language, a 'locus', [29] or, if we think of the hypothetical material structure underlying it, a 'gene'.) Difference of property, to my view, is really the fundamental concept rather than property itself, notwithstanding the apparent linguistic and logical contradiction of this statement. The differences of properties actually are discrete, as will emerge in the next chapter when we have to speak of mutations and the dry scheme hitherto presented will, as I hope, acquire more life and colour.

当然，这里描绘的遗传机制的方案还是相当空洞而乏味的，甚至略带些肤浅。因为我们并没有说明白，我们所说的性状究竟是什么。

我们把一个本质上是统一“整体”的有机体，分割成互不相同的“性状”，这看来既是不妥当的，也是不可能的。现在，在特定的情况下，我们实际上陈述的是：一对祖先如果在某个特定的方面确实存在着明显的差异（比如，一个是蓝色眼睛，另一个是棕色眼睛），那么，他们的后代不是遗传了这一个，就是遗传了另一个。我们在染色体上确定的位置就是这种差异的所在（专业术语为“位点”，如果考虑到其背后假设的物质结构，也可称之为“基因”）。

我认为，真正的基本概念是性状的差别，而不是性状本身，尽管这样的说法在语言上和逻辑上有着明显的矛盾。性状的差别实际上是相互独立的，我们在下一章讲述突变时还会说到这一点。我希望，迄今所提到的枯燥乏味的模式，都将在接下来的章节中变得更为生动、更为多彩。

MAXIMUM SIZE OF A GENE

We have just introduced the term gene for the hypothetical material carrier of a definite hereditary feature. We must now stress two points which will be highly relevant to our investigation. The first is the size—or, better, the maximum size—of such a carrier; in other words, to how small a volume can we trace the location? The second point will be the permanence of a gene, to be inferred from the durability of the hereditary pattern.

As regards the size, there are two entirely independent estimates, one resting on genetic evidence (breeding experiments), the other on cytological evidence (direct microscopic inspection). The first is, in principle, simple enough. After having, in the way described above, located in the chromosome a considerable number of different (large-scale) features (say of the *Drosophila* fly) within a particular one of its chromosomes, to get the required estimate we need only divide the measured length of that chromosome by the number of features and multiply by the cross-section. For, of course, we count as different only such features as are occasionally separated by crossing-over, so that they cannot be due to the same (microscopic or molecular) structure. On the other hand, it is clear that our estimate can only give a maximum size, because the number of features isolated by genetic analysis is continually increasing as work goes on.

最大的基因

我们刚刚介绍了基因这个名词，基因是假设的物质载体，决定一个特定的遗传性状。现在我们要着重强调两点，这都与我们所有的研究高度相关。第一点，这样一种载体的大小——或者更确切地说，这个载体到底能大到什么程度，换句话说，它是如此之小，我们能否确定它的位置？第二点是基因的稳定性，要从遗传模式的持续时间来推断。

关于基因的大小，有两种完全不同的评估方法。一是根据遗传证据（育种实验），二是细胞学的证据（显微镜的直接观察）。原则上，第一种估计是很简单的。按照上面描述的方法，把一条特定的染色体上的数目巨大而又不同的（宏观的）性状（就以果蝇为例）定位到染色体上以后，将测量到的那条染色体的长度除以性状的数目，再乘以染色体的横截面，就得出了我们所需要的估计数。当然，仅仅是被偶然交换而分离的那部分性状，才被我们看作是不同的性状，所以它们不可能是源于相同的（微观或分子）结构。另一方面，很明显，我们的估计数只能得出最大的体积，这是因为通过遗传学分析而分离的那些性状数目，将随研究的进展而不断增加。

The other estimate, though based on microscopic inspection, is really far less direct. Certain cells of *Drosophila* (namely, those 30 of its salivary glands) are, for some reason, enormously enlarged, and so are their chromosomes. In them you distinguish a crowded pattern of transverse dark bands across the fibre. C. D. Darlington has remarked that the number of these bands (2,000 in the case he uses) is, though, considerably larger, yet roughly of the same order of magnitude as the number of genes located in that chromosome by breeding experiments. He inclines to regard these bands as indicating the actual genes (or separations of genes). Dividing the length of the chromosome, measured in a normal-sized cell by their number (2,000), he finds the volume of a gene equal to a cube of edge 300 Å. Considering the roughness of the estimates, we may regard this to be also the size obtained by the first method.

另一种估计，尽管是基于显微镜的观察，但实际上根本不是直接的估算。果蝇的某些细胞（即它的那些唾液腺细胞）由于某种机制变得很大，它们的染色体也是如此。在这些染色体上，你可以清晰地分辨出染色体纤丝上的密集的横向暗带。达林顿就曾经说过，这些条带的数目（在他使用的例子中为2,000条）尽管很大，但大体上与用育种实验得出的基因数目在同一个数量级上。他倾向于相信这些条带标明了实际的基因（或基因的分离）。在一个正常大小的细胞里测得的染色体长度，除以条带的数目（2,000），他所得出的一个基因的体积等于边长为300埃的立方体的体积。考虑到这种方法对基因数目的估计太过粗糙，我们姑且认为这种方法跟第一种算出的体积相差无几。

SMALL NUMBERS

A full discussion of the bearing of statistical physics on all the facts I am recalling—or perhaps, I ought to say, of the bearing of these facts on the use of statistical physics in the living cell—will follow later. But let me draw attention at this point to the fact that 300 Å is only about 100 or 150 atomic distances in a liquid or in a solid, so that a gene contains certainly not more than about a million or a few million atoms. That number is much too small (from the $\sqrt{n}$ point of view) to entail an orderly and lawful behaviour according to statistical physics—and that means according to physics. It is too small, even if all these atoms played the same role, as they do in a gas or in a drop of liquid. And the gene is most certainly not just a homogeneous drop of liquid. It is probably a large protein molecule, in which every atom, every radical, every heterocyclic ring plays an individual role, more or less different from that played by any of the other similar atoms, radicals, or rings. This, at any rate, is the opinion of leading geneticists such as Haldane and Darlington, and we shall soon have to refer to genetic experiments which come very near to proving it.

PERMANENCE

31 Let us now turn to the second highly relevant question: What degree of permanence do we encounter in hereditary properties and

最小的数目

统计物理学对我所能回忆的所有事实的影响 —— 也许，我应该说，这些实例对在活细胞中使用统计物理的影响将在下文中阐述。不过我注意到一个事实：300埃在液体或固体中大约只有100个或150个原子距离。所以，一个基因所含的原子，肯定不会超过一百万或几百万个。遵循统计物理学，这个数字实在太小了（从$\sqrt{n}$的观点来看），不能在物理学上产生有秩序、有规律的行为。它太小了，即使所有这些原子都起相同的作用，就像它们在一团气体或在一滴液体中那样。而且，基因肯定不只是一滴同质的液体。它也许是一个蛋白质大分子，其中每一个原子、每一个基团、每一个杂合闭环都各司其职，多少不同于任何一个相似的原子、基团或闭环所起的作用。不管怎么说，这是引领这个领域的遗传学家霍尔丹和达林顿的意见，而且我们接下来将用遗传学实验对这些观点进行论证。

遗传的稳定性

现在，让我们谈谈第二个高度相关的问题：遗传性状的稳定性有多强？有多少是因为承载这些遗传特性的物质结构？

what must we therefore attribute to the material structures which carry them?

The answer to this can really be given without any special investigation. The mere fact that we speak of hereditary properties indicates that we recognize the permanence to be almost absolute. For we must not forget that what is passed on by the parent to the child is not just this or that peculiarity, a hooked nose, short fingers, a tendency to rheumatism, haemophilia, dichromasy, etc. Such features we may conveniently select for studying the laws of heredity. But actually it is the whole (four-dimensional) pattern of the 'phenotype', the visible and manifest nature of the individual, which is reproduced without appreciable change for generations, permanent within centuries—though not within tens of thousands of years—and borne at each transmission by the material in a structure of the nuclei of the two cells which unite to form the fertilized egg cell. That is a marvel—than which only one is greater; one that, if intimately connected with it, yet lies on a different plane. I mean the fact that we, whose total being is entirely based on a marvellous interplay of this very kind, yet possess the power of acquiring considerable knowledge about it. I think it possible that this knowledge may advance to little short of a complete understanding—of the first marvel. The second may well be beyond human understanding.

这个问题的答案无须做专门的研究。仅从我们谈到的遗传特性这个事实，就表明我们已经承认了遗传的稳定性几乎是绝对的。因为我们不能忘记，父母遗传给子女的并不仅仅是这种或那种特征，比如鹰钩鼻，短手指，患风湿病、血友病、双色色盲的倾向等，我们可以方便地选择这些特征来研究遗传规律。但实际上，这种特征是整个（四维的）模式的“表现型”，是个体的可见的和明显的特性，它在几代人中重复出现而没有明显的变化，在几个世纪内几乎永恒——尽管不是在几万年内——每次都是由两个细胞的细胞核结构中的物质来传递的，这些细胞结合在一起形成受精卵细胞。这真是个奇迹——只有一个事件比这个奇迹更伟大；如果它同我们所说的奇迹是密切相关的话，那也是在不同层面上的奇迹。我的意思是：我们的全部存在，完全是依靠这个奇迹的这种相互作用，而我们又好像有能力去获得有关这种奇迹的更多知识。我认为，我们是有可能掌握这些知识的，并且这些知识可以引领我们进一步接近第一个奇迹。第二个可能超越了人类的认知范围。

CHAPTER 3
Mutations

Und was in schwankender Erscheinung schwebt,
Befestiget mit dauernden Gedanken.[1]

GOETHE

'JUMP-LIKE' MUTATIONS—THE WORKING-GROUND OF NATURAL SELECTION

32 The general facts which we have just put forward in evidence of the durability claimed for the gene structure, are perhaps too familiar to us to be striking or to be regarded as convincing. Here, for once, the common saying that exceptions prove the rule is actually true. If there were no exceptions to the likeness between children and parents, we should have been deprived not only of all those beautiful experiments which have revealed to us the detailed mechanism of heredity, but also of that grand, million-fold experiment of Nature, which forges the species by natural selection and survival of the fittest.

1. And what in fluctuating appearance hovers, Ye shall fix by lasting thoughts.

第三章
突变

常彷徨，始翱翔。

长沉思，方知晓。

——歌德

跳跃式突变——自然选择作用的基础

我们刚才列举的诸多事实，为基因结构的持久性提供了证据。也许对我们来说这些事实已过于熟悉，以至于无法引人注目或具有说服力。这一次，倒真如俗话所说：“例外证明了规律的真实性。”如果子女和父母之间的相似没有例外，我们也就不会有所有的精彩的实验，从而揭示出详尽的遗传机制，大自然本身也就无法通过自然选择和适者生存，从而以几百万倍规模的实验，创造出多种多样的物种。

Let me take this last important subject as the starting-point for presenting the relevant facts—again with an apology and a reminder that I am not a biologist:

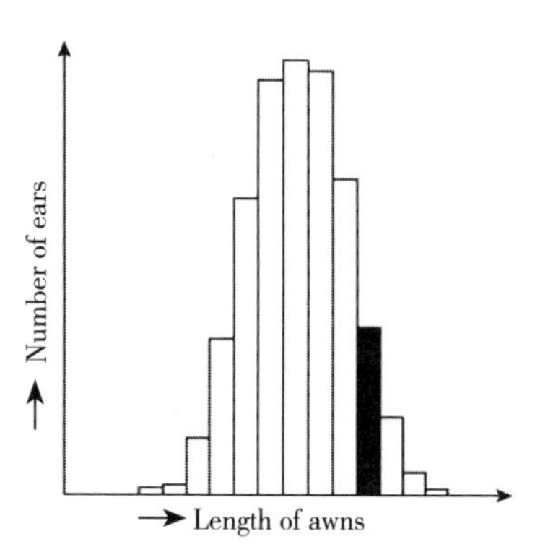

Fig. 7. Statistics of length of awns in a pure-bred crop. The black group is to be selected for sowing. (The details are not from an actual experiment, but are just set up for illustration.)

We know definitely, today, that Darwin was mistaken in regarding the small, continuous, accidental variations, that are bound to occur even in the most homogeneous population, as the material on which natural selection works. For it has been proved that they are not inherited. The fact is important enough to be illustrated brief-
33 ly. If you take a crop of pure-strain barley, and measure, ear by ear, the length of its awns and plot the result of your statistics, you will get a bell-shaped curve as shown in Fig. 7, where the number of ears with a definite length of awn is plotted against the length. In other words: a definite medium length prevails, and deviations in either direction occur with certain frequencies.

请允许我从最后这一重要的主题出发，来陈述相关事实——我想再次深表歉意，并提醒大家我不是一个生物学家。

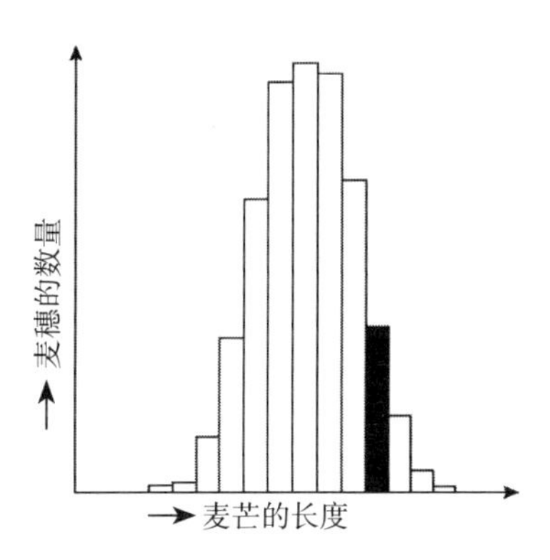

图7　纯系大麦中麦芒长度的统计。涂黑的那一组被选来播种
（具体细节并不是根据实际实验得出，仅作说明之用）

今天，我们十分肯定，达尔文错误地把微小的、连续的、偶然的变异，这些即使在最同质的种群中也必然会发生的变异视为自然选择起作用的物质基础。但现已证明这种变异是不遗传的。这个事实非常重要，值得简要叙述。如果你挑选大麦植株，一穗一穗地逐个测量麦穗的麦芒长度，并把你的统计结果制作成图，你会得到一条钟形曲线，如图7所示。图中，纵轴是具有某一特定长度的麦芒的麦穗数量，横轴是麦芒的长度。也就是说：具有中等长度麦芒的麦穗肯定最多，而更长或者更短的麦芒也有一定的概率。

Now pick out a group of ears (as indicated by blackening) with awns noticeably beyond the average, but sufficient in number to be sown in a field by themselves and give a new crop. In making the same statistics for this, Darwin would have expected to find the corresponding curve shifted to the right. In other words, he would have expected to produce by selection an increase of the average length of the awns. That is not the case, if a truly pure-bred strain of barley has been used. The new statistical curve, obtained from the selected crop, is identical with the first one, and the same would be the case if ears with particularly short awns had been selected for seed. Selection has no effect—because the small, continuous variations are not inherited. They are obviously not based on the structure of the hereditary substance, they are accidental. But about forty years ago 34 the Dutchman de Vries discovered that in the offspring even of thoroughly pure-bred stocks, a very small number of individuals, say two or three in tens of thousands, turn up with small but 'jump-like' changes, the expression 'jump-like' not meaning that the change is so very considerable, but that there is a discontinuity inasmuch as there are no intermediate forms between the unchanged and the few changed. De Vries called that a mutation. The significant fact is the discontinuity. It reminds a physicist of quantum theory—no intermediate energies occurring between two neighbouring energy levels. He would be inclined to call de Vries's mutation theory, figuratively, the quantum theory of biology. We shall see later that this is much more than figurative. The mutations are actually due to quantum jumps in the gene molecule. But quantum theory was but two years old when de Vries first published his discovery, in 1902. Small wonder that it took another generation to discover the intimate connection!

现在，挑选一组麦穗（如涂黑所示），其芒的长度明显高于平均值，将这组足够数量的麦穗播种在麦田里，长成新的庄稼。使用同样的方法对新植株做统计，达尔文也许会期待新的钟形曲线右移。也就是说，达尔文的理论认为，通过选择，可以增加麦芒的平均长度。但事实并非如此，如果用的是真正纯种的大麦品系。根据所选作物绘制成的新的统计曲线，和上一代是完全一样的。如果选择麦芒明显较短的麦穗进行播种，结果也还是如此。选择不会产生影响——因为这种微小的、连续的变异并不会遗传。这些变异显然并不依赖于遗传物质的结构，它们完全是偶然性的。

但是，约40年前，荷兰人德弗里斯发现，即使是完全纯种的植株产生的后代里，其中也有一小部分个体，比如说大约万分之二三，表现出微小"跳跃式"的变化。这里对"跳跃式"的表述并不是说变化非常显著，而是指这一变化存在不连续性，因为在没发生变化的后代和少数发生变化的后代中间，没有过渡的中间状态。德弗里斯称这种变化为突变。它的重要性质是非连续性。这让物理学家联想到量子理论——在两个相邻的能级之间，没有过渡的能级。物理学家倾向于把德弗里斯的突变理论形象地比作生物学中的量子理论。之后我们将看到，这不仅仅是一种比喻。突变实际上就是由基因分子的量子跃迁产生的。但是，德弗里斯首次发表他的理论是在1902年，那时，量子理论才面世2年。难怪又过了一代人的时间，人们才发现这两者之间的密切联系！

THEY BREED TRUE, THAT IS, THEY ARE PERFECTLY INHERITIED

Mutations are inherited as perfectly as the original, correctly unchanged characters were. To give an example, in the first crop of barley considered above a few ears might turn up with awns considerably outside the range of variability shown in Fig. 7, say with no awns at all. They might represent a de Vries mutation and would then breed perfectly true, that is to say, all their descendants would be equally awnless.

Hence a mutation is definitely a change in the hereditary treasure and has to be accounted for by some change in the hereditary substance. Actually most of the important breeding experiments, which have revealed to us the mechanism of heredity, consisted in a careful analysis of the offspring obtained by crossing, according to a preconceived plan, mutated (or, in many cases, multiply mutated) with non-mutated or with differently mutated individuals. On the other hand, by virtue of their breeding true, mutations are a suitable 35 material on which natural selection may work and produce the species as described by Darwin, by eliminating the unfit and letting the fittest survive. In Darwin's theory, you just have to substitute 'mutations' for his 'slight accidental variations' (just as quantum theory substitutes 'quantum jump' for 'continuous transfer of energy'). In all other respects little change was necessary in Darwin's theory,

纯系的完美遗传

突变可以完美地遗传，准确地保持祖先的性状而没有改变。举个例子，在上述提到的第一代大麦中，也许会出现几个麦穗，其芒的长度大大超过图7所示的变化范围，或许根本没有麦芒。这也许就是德弗里斯突变，如果可以进行纯种选育，我们不得不说，以它培育出的所有后代都是无芒的。

因此，突变肯定是遗传体系之内发生的不可多得的变化，而且一定可以解释为遗传物质发生了某些变化。事实上，大多数重要的育种实验，已经向我们揭示了遗传机制，这些实验按照设定的计划，一方面将突变的（或在很多情况下，发生的是多重突变）个体和未突变的，或者发生不同突变的个体进行杂交，然后仔细分析杂交所获得的子代个体；另一方面，由于突变可以进行纯种繁殖，突变也成了自然选择作用的物质基础，按照达尔文适者生存不适者淘汰的原则，从而孕育了物种。

根据达尔文的理论，你只需用“突变”来替代他所说的“微小的偶然变异”即可（就像在量子理论中用“量子跃迁”代替“能

that is, if I am correctly interpreting the view held by the majority of biologists.[1]

LOCALIZATION, RECESSIVITY AND DOMINANCE

We must now review some other fundamental facts and notions about mutations, again in a slightly dogmatic manner, without showing directly how they spring, one by one, from experimental evidence.

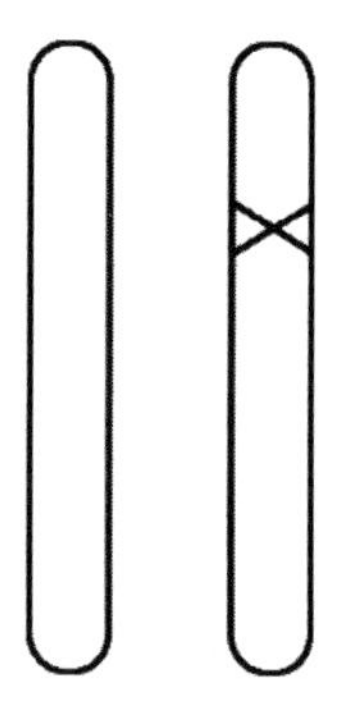

Fig. 8. Heterozygous mutant. The cross marks the mutated gene.

1. Ample discussion has been given to the question, whether natural selection be aided (if not superseded) by a marked inclination of mutations to take place in a useful or favourable direction. My personal view about this is of no moment; but it is necessary to state that the eventuality of 'directed mutations' has been disregarded in all the following. Moreover, I cannot enter here on the interplay of 'switch' genes and 'polygenes', however important it be for the actual mechanism of selection and evolution.

量的连续转移”那样）。在其他所有方面，达尔文理论几乎没必要改动，如果我在正确地转述大多数生物学家所持的观点。[1]

基因的定位，隐性和显性

现在我们必须以一种不那么教条的方式回顾一些有关突变的其他基本事实和概念，而不是直接地展示它们是如何一个接一个地不断出现在实验证据之中。

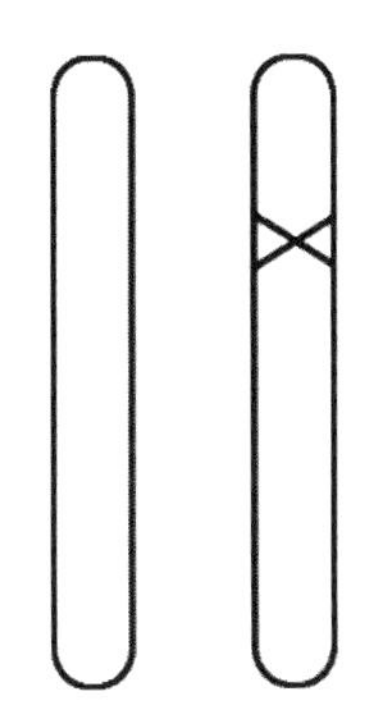

图8　杂合的突变体。× 表示突变的位点

1. 这一问题已有很多讨论，朝着有益的或有利的方向发展的明显的变异倾向是否有助于(如果不是替代)自然选择。我个人关于这个问题的看法并不重要，但是有必要说明，大家似乎忽视了“定向突变”的可能性。我这儿不对“开关基因”和“多基因”的相互作用做过多的阐述，虽然这种作用对于实际的选择和演化机制非常重要。

We should expect a definite observed mutation to be caused by 36 a change in a definite region in one of the chromosomes. And so it is. It is important to state that we know definitely that it is a change in one chromosome only, but not in the corresponding 'locus' of the homologous chromosome. Fig. 8 indicates this schematically, the cross denoting the mutated locus. The fact that only one chromosome is affected is revealed when the mutated individual (often called 'mutant') is crossed with a non-mutated one. For exactly half of the offspring exhibit the mutant character and half the normal one. That is what is to be expected as a consequence of the separation of the two chromosomes on meiosis in the mutant—as shown, very schematically, in Fig. 9. This is a 'pedigree', representing every individual (of three consecutive generations) simply by the pair of chromosomes in question. Please realize that if the mutant had both its chromosomes affected, all the children would receive the same (mixed) inheritance, different from that of either parent.

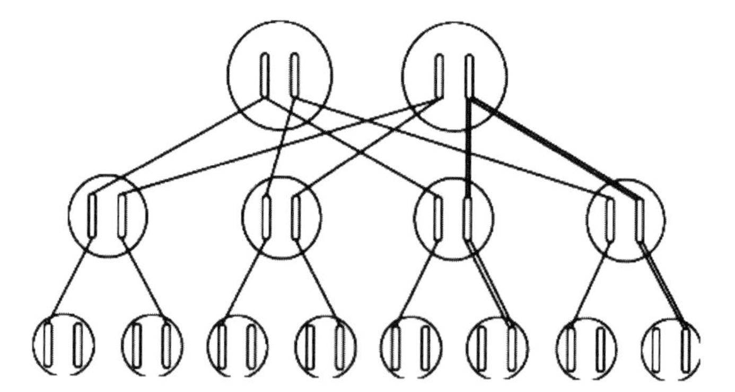

Fig. 9. Inheritance of a mutation. The straight lines across indicate the transfer of a chromosome, the double ones that of the mutated chromosome. The unaccounted-for chromosomes of the third generation come from the *mates* of the second generation, which are not included in the diagram. They are supposed to be non-relatives, free of the mutation.

我们应该期望一个明确观察到的突变是由于一条染色体的某一区段内发生了改变所引起的。事实确实如此。更为重要的是，我们肯定知道，这只是一条染色体中的一个变化，而没有发生在同源染色体对应的“位点”上。图8展示了这一点，这里的X表示突变的位点。事实表明，当突变的个体（常称为“突变体”）同一个非突变个体杂交时，只有一条染色体受到影响。因为子代中正好有一半显示突变体的性状，另一半则是正常的。这正好与突变体减数分裂时两条同源染色体分离的现象一致，如图9所示。这是一个“谱系”，用所讨论的一对染色体来表示每个个体（连续三代），要知道，如果突变体的两条染色体都受到影响，那么，它的所有子女都会遗传到同一种（混合的）性状，这种遗传性状既不同于他们的父亲，也不同于他们的母亲。

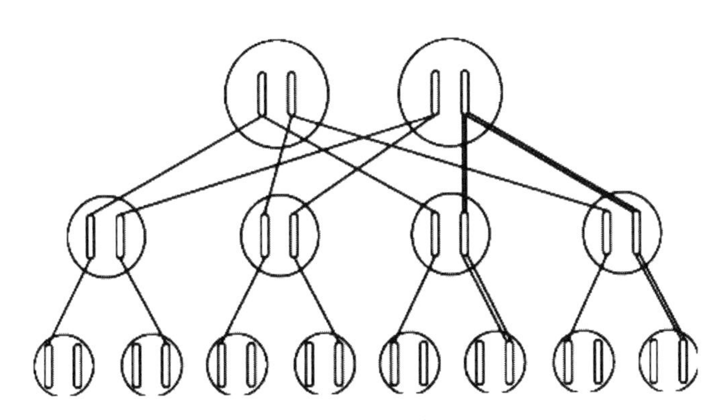

图9　突变的遗传

图中交叉的直线表示某条染色体的转移，其中双线表示发生突变的染色体。第三代中所获得的未突变的染色体来自第二代中相应的配偶，未在图中标出。假定这些配偶非亲缘关系，也无突变发生。

But experimenting in this domain is not as simple as would appear from what has just been said. It is complicated by the second important fact, viz. that mutations are very often latent. What does that mean?

37

In the mutant the two ‘copies of the code-script’ are no longer identical; they present two different ‘readings’ or ‘versions’, at any rate in that one place. Perhaps it is well to point out at once that, while it might be tempting, it would nevertheless be entirely wrong to regard the original version as ‘orthodox’, and the mutant version as ‘heretic’. We have to regard them, in principle, as being of equal right—for the normal characters have also arisen from mutations.

可是，这个领域里的实验可不像刚才所说的那么简单。第二个重要的事实使得研究变得复杂了，突变往往是藏而不露的，这是什么意思呢？

在突变体中，两份“密码本的拷贝”不再是一模一样的。在同一处，不管如何，它们呈现的是两个不同的“解读”或“版本”。也许应该马上指出，尽管这很诱人，但把原始的版本看作是“正统的”，把突变体的版本看成是“异端的”，那完全是错误的。从原则上讲，我们必须认同它们是有同等权利的——因为正常性状也是起源于突变。

What actually happens is that the 'pattern' of the individual, as a general rule, follows either the one or the other version, which may be the normal or the mutant one. The version which is followed is called dominant, the other, recessive; in other words, the mutation is called dominant or recessive, according to whether it is immediately effective in changing the pattern or not.

Recessive mutations are even more frequent than dominant ones and are very important, though at first they do not show up at all. To affect the pattern, they have to be present in both chromosomes (see Fig. 10). Such individuals can be produced when two equal recessive mutants happen to be crossed with each other or when a mutant is crossed with itself; this is possible in hermaphroditic plants and even happens spontaneously. An easy reflection shows that in these cases about one-quarter of the offspring will be of this type and thus visibly exhibit the mutated pattern.

38

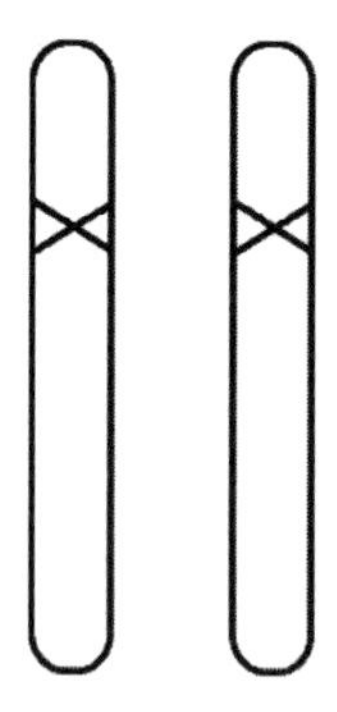

Fig. 10. Homozygous mutant, obtained in one-quarter of the descendants either from self-fertilization of a heterozygous mutant (see Fig. 8) or from crossing two of them.

一般说来，实际发生的情况是，根据一般的定律，个体的“性状”如果不是与这个版本一样，便是与另一个版本一样。这些版本可以是正常的，也可以是突变的。随后，一个版本称为显性，另一个则称为隐性。换句话说，这种突变称为显性突变还是隐性突变，取决于它是否直接影响到性状的改变。

隐性突变甚至比显性突变更为常见且十分重要，尽管刚开始它的性状一点也不表现出来。为了影响到性状，它们一定要在两条染色体上都出现（见图10）。两个一样的隐性突变体一旦杂交，或一个突变体自交时，就能产生这样的个体。这可能发生在雌雄同株的植物里，甚至可以自发产生。只要稍微注意一下就可看到，在这种情况下，子代中约有四分之一是属于这种类型的，而且明显地表现出突变了的性状。

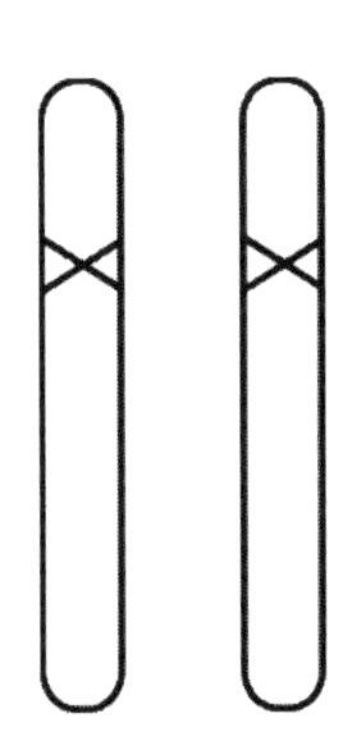

图10　突变体的纯合子，一个突变体的杂合子进行自交（见图8）或者两个突变体的杂合子进行杂交后，四分之一的后代为突变体纯合子。

INTRODUCING SOME TECHNICAL LANGUAGE

I think it will make for clarity to explain here a few technical terms. For what I called 'version of the code-script'—be it the original one or a mutant one—the term 'allele' has been adopted. When the versions are different, as indicated in Fig. 8, the individual is called heterozygous, with respect to that locus. When they are equal, as in the non-mutated individual or in the case of Fig. 10, they are called homozygous. Thus a recessive allele influences the pattern only when homozygous, whereas a dominant allele produces the same pattern, whether homozygous or only heterozygous.

Colour is very often dominant over lack of colour (or white). Thus, for example, a pea will flower white only when it has the 'recessive allele responsible for white' in both chromosomes in question, when it is 'homozygous for white'; it will then breed true, and all its descendants will be white. But one 'red allele' (the other being white; 'heterozygous') will make it flower red, and so will two red alleles ('homozygous'). The difference of the latter two cases will only show up in the offspring, when the heterozygous red will produce some white descendants, and the homozygous red will breed true.

The fact that two individuals may be exactly alike in their outward appearance, yet differ in their inheritance, is so important that an exact differentiation is desirable. The geneticist says they have the same phenotype, but different genotype. The contents of the pre-

一些新的术语

在这里我想解释一些专业术语，来把问题讲清楚。当我讲到“密码本”，不管是原始版本还是突变版本，实际上已经采用了“等位基因”这一术语。当两个版本不同时，如图8所示，就那个位点而言，我们就说此个体是杂合子。反之，如果两个版本是同样的，如图10所示的非突变个体，则称之为纯合子。这样，只有在纯合的时候，隐性的等位基因才会影响性状；而显性的等位基因都产生相同的性状，不管是在纯合还是杂合的个体中。

有色对于没有颜色（或白色）来讲，往往是显性的。比如，开白花的豌豆只有在它的两个相关的染色体都存在着“负责白色的隐性等位基因”时，也就是在“白色纯合子”的时候，才会开白花，这时候才能繁育纯系，并且所有子代全都是开白花的。但是，如果有一个“红色等位基因”（另一个等位基因是白色的，个体就是“杂合子”），豌豆就会开红花；有两个红色等位基因（“纯合子”），豌豆也会开红花。这两种情况的区别在子代中才会显示出来，因为杂合的红花会产生一些开白花的子代，而纯合的红花将只产生开红花的子代。

两个个体可能在外观上十分相似，但它们的遗传特性却不相同，这个事实是非常重要的，所以需要精确地予以区分。遗传

ceding paragraphs could thus be summarized in the brief, but highly technical statement:

A recessive allele influences the phenotype only when the genotype is homozygous.

We shall use these technical expressions occasionally, but shall recall their meaning to the reader where necessary.

学家称它们具有相同的表现型，但具有不同的基因型。因此，前面几节的内容可以简短但又非常专业地归纳为：

只有基因型是纯合的时候，隐性等位基因才能影响表现型。

我们不时地会用到这些专业术语，必要时将再向读者说明其含义。

THE HARMFUL EFFECT OF CLOSE-BREEDING

39 Recessive mutations, as long as they are only heterozygous, are of course no working-ground for natural selection. If they are detrimental, as mutations very often are, they will nevertheless not be eliminated, because they are latent. Hence quite a host of unfavourable mutations may accumulate and do no immediate damage. But they are, of course, transmitted to half of the offspring, and that has an important application to man, cattle, poultry or any other species, the good physical qualities of which are of immediate concern to us. In Fig. 9 it is assumed that a male individual (say, for concreteness, myself) carries such a recessive detrimental mutation heterozygously, so that it does not show up. Assume that my wife is free of it. Then half of our children (second line) will also carry it—again heterozygously. If all of them are again mated with non-mutated partners (omitted from the diagram, to avoid confusion), a quarter of our grandchildren, on the average, will be affected in the same way.

No danger of the evil ever becoming manifest arises, unless equally affected individuals are crossed with each other, when, as an easy reflection shows, one-quarter of their children, being homozygous, would manifest the damage. Next to self-fertilization (only possible in hermaphrodite plants) the greatest danger would be a marriage between a son and a daughter of mine. Each of them standing an even chance of being latently affected or not, one-quarter of these incestuous（乱伦的）unions would be dangerous inasmuch as one-quarter of its children would manifest the damage. The danger factor for an incestuously bred child is thus 1 ∶ 16.

近亲繁殖的恶果

隐性突变只要还是杂合的，自然选择对它们当然就没有了作用的基础。如果突变是有害的，确实常常如此，它们也不会被剔除，因为它们隐而不见。因此，大量的不利突变可能会累积起来，而不会立即产生危害。但它们一定会传递给子代中的半数个体，这对人类、家畜、家禽或任何其他物种都有重要的应用，这些物种的良好体质是我们直接关心的。

在图9中，假定一个男性个体（说具体些，比如我自己），携带这样杂合的隐性有害突变，它并没有表现出来。假如我的夫人没有这种突变，那么我的子女中将有半数（图中的第二行）也会带有这种突变，而且也是杂合的。倘若他们全部再与不携带突变基因的配偶结婚（图中略去了子女的配偶，以免混淆），那么，在我们的孙辈中，平均将会有四分之一获得这种隐性突变。

除非与有同样突变的个体婚配，否则有害的性状永远不会显现出来，这时，只要稍作考虑就可看到，当他们的子女中有四分之一是纯合的时候，就会表现出这种危害。除了自体受精（实际上，只有雌雄同株的植物才有此可能）之外，危害最大的是我的一个儿子同我的一个女儿结婚。他们中的每一个人，受或不受潜在危害影响的机会是相等的，这种乱伦的结合中有四分之一的后代是危险的，因为他们的子女中间有四分之一将受影响。因此，对乱伦生下来的孩子来说，危险系数为1∶16。

In the same way the danger factor works out to be 1 ∶ 64 for the offspring of a union between two ('clean-bred') grandchildren of mine who are first cousins. These do not seem to be overwhelming odds, and actually the second case is usually tolerated. But do not forget that we have analysed the consequences of only one possible latent injury in one partner of the ancestral couple ('me and my wife'). Actually both of them are quite likely to harbour more than one latent deficiency of this kind. If you know that you yourself harbour a definite one, you have to reckon with 1 out of 8 of your first cousins sharing it! Experiments with plants and animals seem to indicate that in addition to comparatively rare deficiencies of a serious kind, there seem to be a host of minor ones whose chances combine to deteriorate the offspring of close-breeding as a whole. Since we are no longer inclined to eliminate failures in the harsh way the Lacedaemonians used to adopt in the Taygetos mountain, we have to take a particularly serious view about these things in the case of man, where natural selection of the fittest is largely retrenched, nay, turned to the contrary. The anti-selective effect of the modern mass slaughter of the healthy youth of all nations is hardly outweighed by the consideration that in more primitive conditions war may have had a positive value in letting the fittest tribe survive.

同样危险的是：我的两个（“纯血缘的”）孙子孙女，即堂、表兄妹之间结婚生下的后代的风险系数是1∶64。这种概率看上去并不太大，事实上，第二种情况通常是可以接受的。可是不要忘了我们已经分析过的，上代夫妇（“我和我的妻子”）中，一方携带了一个可能造成危害的突变。事实上，他们两人携带的这种潜在的缺陷很可能都不止一个。如果你知道自己肯定携带了一个缺陷，那么，就可以推算出，在你的8个堂、表兄妹中，有一个也会带有这种缺陷。用动植物进行的实验似乎表明，除了一些严重的、比较罕见的缺陷外，还有很多较小的缺陷，产生这些缺陷的概率加在一起，就会使得整个近亲繁殖的后代衰退恶化。

既然如今我们不再用斯巴达人在泰杰托斯山经常采用的那种残暴方式去消灭弱势者，那么，在人类中，自然的适者生存的选择作用大大减弱了，不！是径直走向对立面，我们必须严肃地看待在人类中发生的这些事情。在更原始的条件下，战争也许对选择出最合适生存的部落有积极的价值；而现代大量屠杀所有国家的健康青年的反选择效应，连这一点理由也不存在了。

GENERAL AND HISTORICAL REMARKS

The fact that the recessive allele, when heterozygous, is completely overpowered by the dominant and produces no visible effects at all, is amazing. It ought at least to be mentioned that there are exceptions to this behaviour. When homozygous white snapdragon（金鱼草）is crossed with, equally homozygous, crimson snapdragon, all the immediate descendants are intermediate in colour, i.e. they are pink (not crimson, as might be expected). A much more important case of two alleles exhibiting their influence simultaneously occurs in blood-groups—but we cannot enter into that here. I should not be astonished if at long last recessivity should turn out to be capable of degrees and to depend on the sensitivity of the tests we apply to examine the 'phenotype'.

This is perhaps the place for a word on the early history of genetics. The backbone of the theory, the law of inheritance, to successive generations, of properties in which the parents differ, and more especially the important distinction recessive-dominant, are due to the now world famous Augustininan Abbot Gregor Mendel (1822—41 1884). Mendel knew nothing about mutations and chromosomes. In his cloister（修道院）gardens in Brünn (Brno) he made experiments on the garden pea, of which he reared different varieties, crossing them and watching their offspring in the 1st, 2nd, 3rd, ⋯, generation. You might say, he experimented with mutants which he found ready-made in nature. The results he published as early as 1866 in

普遍性和历史观

事实是，隐性等位基因在杂合时完全被显性等位基因所掩盖，丝毫不产生可观察到的效应，这是令人惊奇的。不过应该说还是有不少例外，比方说，当纯合的白色金鱼草与同样是纯合的深红色的金鱼草杂交时，所有的直系后代都是中间色的，即粉红色的（而不是预期的深红色的）。更能说明问题的例子是血型，两个等位基因同时显示出它们的影响——不过我们不展开讨论这个话题了。我对此并不会感到惊讶，如果最终能够证明隐性有不同程度之分，并且取决于我们用来检测“表现型”的实验的灵敏度。

这里，也许要讲几句遗传学的早期历史。这个理论的主体，即遗传定律，对连续几代人来说，父母不同的特性，尤其是关于显-隐性之间的重要区别，都应归功于如今享誉全球的奥古斯汀修道士格雷戈尔·孟德尔（1822—1884）。一开始，孟德尔对突变与染色体一无所知。在布隆（布鲁诺）修道院花园中，他用豌豆做实验。他培育了很多不同品种的豌豆，让它们进行杂交，并注意观察它们的第一代、第二代、第三代……你当然可以说，他用于实验的突变体都是自然界里现成的。早在1866年，

the Proceedings of the *Naturforschender Verein in Brünn*. Nobody seems to have been particularly interested in the abbot's hobby, and nobody, certainly, had the faintest idea that his discovery would in the twentieth century become the lodestar（指导原则）of an entirely new branch of science, easily the most interesting of our days. His paper was forgotten and was only rediscovered in 1900, simultaneously and independently, by Correns (Berlin), de Vries (Amsterdam) and Tschermak (Vienna).

他就把实验结果发表在《布鲁诺自然研究会学报》上。当时，多数人对这个修道士的爱好不屑一顾，确实也没有人会想到，他的发现在20世纪竟会成为一个全新科学领域的灯塔，无疑也是当今最有趣的学科之一。他的论文后来被人遗忘，直到1900年才被科伦斯（柏林）、德弗里斯（阿姆斯特丹）和切尔玛克（维也纳）三人同时又独立地重新发现。

THE NECESSITY OF MUTATION BEING A RARE EVENT

So far we have tended to fix our attention on harmful mutations, which may be the more numerous; but it must be definitely stated that we do encounter advantageous mutations as well. If a spontaneous mutation is a small step in the development of the species, we get the impression that some change is 'tried out' in rather a haphazard fashion at the risk of its being injurious, in which case it is automatically eliminated. This brings out one very important point. In order to be suitable material for the work of natural selection, mutations must be rare events, as they actually are. If they were so frequent that there was a considerable chance of, say, a dozen of different mutations occurring in the same individual, the injurious ones would, as a rule, predominate over the advantageous ones and the species, instead of being improved by selection, would remain unimproved, or would perish. The comparative conservatism which results from the high degree of permanence of the genes is essential. An analogy might be sought in the working of a large manufacturing plant in a factory. For developing better methods, innovations, even 42 if as yet unproved, must be tried out. But in order to ascertain whether the innovations improve or decrease the output, it is essential that they should be introduced one at a time, while all the other parts of the mechanism are kept constant.

突变只能是罕见事件

迄今为止，我们都试图把注意力集中在有害突变上，因为这种突变也许更为多见；但需要明确指出的是，我们的确也碰到过一些有益的突变。如果一个自发突变是一个物种演化过程中的一小步，那么，我们会得到这样的印象：有些变化是以看似杂乱随意的方式来“尝试”，冒着可能有害而被自动剔除的风险。

由此引出了至关重要的一点，要成为自然选择的合适材料，突变一定是罕见的，事实也是如此。如果突变非常频繁，就会导致同一个体内出现好几十个不同的突变，有害的突变又通常比有利突变更占优势。那么，按照常规，这个物种非但不会通过自然选择而得到进化，反而是没有任何改变，甚至是灭亡。基因高度的持久性导致相对保守的变化，这是极其关键的。

我们可以从大型制造工厂的工作中找到类比，一个工厂为了开发更好的技术，即使大胆的革新还没有得到证实，也是要付之一试的。可是，为了确定这些创新究竟是引起增产还是减产，有必要在一段时间内只采用一项创新，在此期间，该厂的其余部分仍保持不变。

MUTATIONS INDUCED BY X-RAYS

We now have to review a most ingenious series of genetical research work, which will prove to be the most relevant feature of our analysis.

The percentage of mutations in the offspring, the so-called mutation rate, can be increased to a high multiple of the small natural mutation rate by irradiating the parents with X-rays or γ-rays. The mutations produced in this way differ in no way (except by being more numerous) from those occurring spontaneously, and one has the impression that every 'natural' mutation can also be induced by X-rays. In *Drosophila* many special mutations recur spontaneously again and again in the vast cultures; they have been located in the chromosome, as described on pp. 26–9, and have been given special names. There have been found even what are called 'multiple alleles', that is to say, two or more different 'versions' and 'readings'—in addition to the normal, non-mutated one—of the same place in the chromosome code; that means not only two, but three or more alternatives in that particular one 'locus', any two of which are to each other in the relation 'dominant-recessive' when they occur simultaneously in their corresponding loci of the two homologous chromosomes.

The experiments on X-ray-produced mutations give the impression that every particular 'transition', say from the normal individual to a particular mutant, or conversely, has its individual 'X-ray coefficient', indicating the percentage of the offspring which turns out to have mutated in that particular way, when a unit dosage of X-ray has been applied to the parents, before the offspring was engendered.

X射线的诱变

现在我们来回顾遗传学的一系列最巧妙的研究，这些研究将佐证我们分析的最重要的特征。

如果将亲代暴露于X射线或γ射线之下的话，子代发生突变的百分率，即突变率将成倍高于自然突变率。这种方式诱发的突变与自然突变并没有什么不同（除了突变率较高以外），因而会使人产生这样的印象，即所有的“自然”突变都可由X射线诱发产生。在人工培育的巨大的果蝇种群中，很多特定的突变自发地反复出现；正如第二章第7-8节（原文第26-29页）所述，目前这些自发突变已完成了在果蝇染色体上的定位，并被赋予了特殊的名字。甚至还发现了所谓的“复等位基因”，也就是说，在染色体的同一个位置，除了正常的、没有发生突变的遗传密码以外，还有两种或两种以上不同的“版本”和“解读”，这意味着当两个同源染色体的相应位点同时发生突变时，在那个特定的“位点”上不仅有两个，而是有三个或更多的替换，其中任何两个都具有“显性-隐性”的关系。

X射线诱发的突变实验给人这样的印象：每一个特定的“转换”，例如由正常个体变为一个特定的突变体，或反之，都有它自己独立的“X射线系数”，这表明，如果在尚未产生子代前，单位剂量的X射线就照射在亲代身上，子代会按照特定方式突变的百分率。

FIRST LAW. MUTATION IS A SINGLE EVENT

43 Furthermore, the laws governing the induced mutation rate are extremely simple and extremely illuminating. I follow here the report of N. W. Timoféëff, in *Biological Reviews*, vol. IX, 1934. To a considerable extent it refers to that author's own beautiful work. The first law is

(I) *The increase is exactly proportional to the dosage of rays, so that one can actually speak [as I did] of a coefficient of increase.*

We are so used to simple proportionality that we are liable to underrate the far-reaching consequences of this simple law. To grasp them, we may remember that the price of a commodity, for example, is not always proportional to its amount. In ordinary times a shopkeeper may be so much impressed by your having bought six oranges from him, that, on your deciding to take after all a whole dozen, he may give it to you for less than double the price of the six. In times of scarcity the opposite may happen. In the present case, we conclude that the first half-dosage of radiation, while causing, say, one out of a thousand descendants to mutate, has not influenced the rest at all, either in the way of predisposing them for, or of immunizing them against, mutation. For otherwise the second half-dosage would not cause again just one out of a thousand to mutate. Mutation is thus not an accumulated effect, brought about by consecutive small portions of radiation reinforcing each other. It must consist in some single event occurring in one chromosome during irradiation. What kind of event?

第一定律：突变是单一事件

进一步说，关于诱发突变率的规律极其简单，而且极具启发性。在这里我依据的是提莫菲于1934年在《生物学评论》第9卷上发表的报告。这个报告的很大部分内容来自作者自己的出色工作。第一定律是：

（1）突变率的增高与射线剂量呈严格的正相关，因而大家可以（像我这样）以突变系数来表示。

我们习惯于这样简单的比例关系，以至于很容易低估这一简单定律的深远影响。为了理解这些深远影响，我们可能还记得，例如，一件商品的定价并不都和它的总量呈正相关。在平时，店主可能会因为你从他那里买了6个橘子而对你印象深刻，以至于当你决定拿走整整一打橘子时，他可能会以低于6个橘子价格的两倍卖给你。但是在货源紧张时，则可能会发生相反的情况。

就目前的情况而言，我们可以得出结论：假如一半辐射的剂量导致了，比如说，千分之一的子代个体发生了突变，但是丝毫不影响其余的子代，无论是在使他们易于发生突变的方面，还是在使他们产生对突变的免疫能力方面。否则，另一半的辐射剂量就不能正好诱发千分之一的突变体。因此，突变并不是一个累积效应，不是由连续的小剂量辐射相互增强而累积引起的。它一定是射线辐射导致的染色体上的单一事件。那到底是什么事件呢？

SECOND LAW. LOCALIZATION OF THE EVENT

This is answered by the second law, viz.

(2) *If you vary the quality of the rays (wave-length) within wide limits, from soft X-rays to fairly hard γ-rays, the coefficient remains constant, provided you give the same dosage in so-called r-units*, that is to say, provided you measure the dosage by the total amount of ions produced per unit volume in a suitably chosen standard substance during the time and at the place where the parents are exposed to the rays.

44

As standard substance one chooses air not only for convenience, but also for the reason that organic tissues are composed of elements of the same atomic weight as air. A lower limit for the amount of ionizations or allied processes[1] (excitations) in the tissue is obtained simply by multiplying the number of ionizations in air by the ratio of the densities. It is thus fairly obvious, and is confirmed by a more critical investigation, that the single event, causing a mutation, is just an ionization (or similar process) occurring within some 'critical' volume of the germ cell. What is the size of this critical volume? It can be estimated from the observed mutation rate by a consideration of this kind: if a dosage of 50,000 ions per cm^3 produces a chance of only 1 : 1000 for any particular gamete (that finds itself in the irradiated district) to mutate in that particular way,

1. A lower limit, because these other processes escape the ionization measurement, but may be efficient in producing mutations.

第二定律：突变的定位

这是由第二定律来回答的。

(2) 如果你在一个很大的限制范围内改变射线的性质（波长），即从穿透性弱的软X射线至穿透性强的硬γ射线，（突变）系数仍能保持恒定，只要你以所谓的伦琴单位（*r-units*）给予同样的剂量，就是说，只要你在亲代暴露于射线的具体时间和空间之中选择合适的标准物测量的每个单位体积产生的离子总数。

我们之所以选择空气作为标准物，并不只是为了方便，而是因为有机组织和气体都是由原子重量相同的元素组成的。组织中发生电离或类似过程（激发）作用量的下限[1] 可以用空气中的离子数乘以密度比来求得。因此，结果非常明显，并已得到更为严谨的研究证实，诱发突变的单一事件正是在生殖细胞中某些“临界”体积发生的电离（或相似的过程）。

这个“临界”体积到底有多大？这可以从观察到的突变率来估计。假设我们这样考虑：如果每一立方厘米（cm^3）产生50,000个离子的剂量，可以导致任一配子（置于射线区）以特定方式发生突变的几率只有1：1000，那么这一临界体积，即

1. 这是一个较低的极限，因为这种程度的电离足以产生突变，而不影响其他方面。

we conclude that the critical volume, the 'target' which has to be 'hit' by an ionization for that mutation to occur, is only $\frac{1}{1000}$ of $\frac{1}{50000}$ of a cm^3, that is to say, one fifty-millionth of a cm^3. The numbers are not the right ones, but are used only by way of illustration. In the actual estimate we follow M. Delbrück, in a paper by Delbrück, N.W. Timoféëff and K.G. Zimmer,[1] which will also be the principal source of the theory to be expounded in the following two chapters. He arrives there at a size of only about ten average atomic distances cubed, containing thus only about 10^3 = a thousand atoms. The simplest interpretation of this result is that there is a fair chance of producing that mutation when an ionization (or excitation) occurs not more than about '10 atoms away' from some particular spot in the chromosome. We shall discuss this in more detail presently.

1. *Nachr. a. d. Biologie d. Ges. d. Wiss. Göttingen, 1(1935), 189.*

被离子击中而诱发突变的“靶”的体积只有一个立方厘米的$\frac{1}{50000}$的$\frac{1}{1000}$，即1立方厘米的$\frac{1}{50000000}$。

这个数字并不真的那么准确，只是用来作为一种解释。在实际的估算中，我们遵循的是德尔布吕克在他的一篇与提莫菲、齐默尔合作的论文[1]中的观点，也将是下面两章中阐述的理论的主要来源。他得出，这个体积仅仅只是10个原子距离的立方，即仅含有大约$10^3=1000$个原子。这一结果的最简单的解释是，当电离（激发）发生在染色体上某一特定位置“10个原子”距离以内的地方，就有相当大的机会产生突变。我们将在下文中更详细地讨论这个问题。

1.《生物学》，[*Nachr.a.d.Biologie d. Ges. D. Wiss. Göttingen*]，第189期，1935年。

The Timoféëff report contains a practical hint which I cannot refrain from mentioning here, though it has, of course, no bearing on our present investigation. There are plenty of occasions in modern 45 life when a human being has to be exposed to X-rays. The direct dangers involved, as burns, X-ray cancer, sterilization, are well known, and protection by lead screens, lead-loaded aprons, etc., is provided, especially for nurses and doctors who have to handle the rays regularly. The point is, that even when these imminent dangers to the individual are successfully warded off, there appears to be the indirect danger of small detrimental mutations being produced in the germ cells—mutations of the kind envisaged when we spoke of the unfavourable results of close-breeding. To put it drastically, though perhaps a little naïvely, the injuriousness of a marriage between first cousins might very well be increased by the fact that their grandmother had served for a long period as an X-ray nurse. It is not a point that need worry any individual personally. But any possibility of gradually infecting the human race with unwanted latent mutations ought to be a matter of concern to the community.

提莫菲的报告中含有一个实际的启示，我不得不在此提及，虽然这与我们目前的探讨没有关系。在现代社会很多场合中，一个人不可避免地要暴露在X射线之下。所涉及的直接危险，如灼伤、X射线癌症和绝育，是众所周知的，铅屏或者铅围裙都提供了保护，特别是对于那些经常操作这些射线的护士和医生。问题是，即使是成功地避免了对这些人的直接伤害，生殖细胞中产生有害的小突变的间接危险似乎仍然存在——这即是我们讲的近亲繁殖的不利结果时所设想的那种突变。说得严重点，虽然可能有点天真，但堂亲、表亲之间的婚配带来的伤害的程度也可能因为他（她）的祖母曾长时间从事X射线护士的工作而增加。虽然，这不是任何个人需要担心的问题，但任何慢慢影响人类的不理想的潜在突变都应当受到社会关注。

CHAPTER 4
The Quantum-Mechanical Evidence

Und deines Geistes höchster Feuerflug
Hat schon am Gleichnis, hat am Bild genug.[1]

GOETHE

PERMANENCE UNEXPLAINABLE BY CLASSICAL PHYSICS

46 Thus, aided by the marvellously subtle instrument of X-rays (which, as the physicist remembers, revealed thirty years ago the detailed atomic lattice structures of crystals), the united efforts of biologists and physicists have of late succeeded in reducing the upper limit for the size of the microscopic structure, being responsible for a definite large-scale feature of the individual—the 'size of a gene'—and reducing it far below the estimates obtained on pp. 29–30. We are now seriously faced with the question: How can we, from the point of view of statistical physics, reconcile the facts that the gene structure seems to involve only a comparatively small number of atoms (of the order of 1,000 and possibly much less), and that nevertheless it displays a most regular and lawful activity—with a durability or permanence that borders upon the miraculous?

1. And thy spirit's fiery flight of imagination acquiesces in an image, in a parable.

第四章
量子力学的论据

任思维展开想象的翅膀，
在形象比喻的太空翱翔。

——歌德

经典物理学不能解释基因的稳定性

因此，借助精巧异常的X射线精密仪器（物理学家都还记得，30年前，这一仪器就揭示了晶体的详细晶格结构），经过生物学家和物理学家的共同努力，最终成功地缩小了微观结构大小的上限；确定了个体的宏观特征——“基因的大小”——比之前的估算（见第二章第9节，原文第29-30页）要小得多。而现在我们面临的严肃问题是：从统计物理学的角度来看，基因结构似乎包含相对较少的原子（1,000个的数量级，也可能更小），但却产生了最有秩序、有规律的行为，并奇迹般地具有持久性与稳定性，我们如何协调这两方面的客观事实呢？

Let me throw the truly amazing situation into relief once again. Several members of the Habsburg dynasty have a peculiar disfigurement of the lower lip ('Habsburger Lippe'). Its inheritance has been studied carefully and published, complete with historical portraits, by the Imperial Academy of Vienna, under the auspices of the 47 family. The feature proves to be a genuinely Mendelian 'allele' to the normal form of the lip. Fixing our attention on the portraits of a member of the family in the sixteenth century and of his descendant, living in the nineteenth, we may safely assume that the material gene structure, responsible for the abnormal feature, has been carried on from generation to generation through the centuries, faithfully reproduced at every one of the not very numerous cell divisions that lie between. Moreover, the number of atoms involved in the responsible gene structure is likely to be of the same order of magnitude as in the cases tested by X-rays. The gene has been kept at a temperature around 98°F during all that time. How are we to understand that it has remained unperturbed by the disordering tendency of the heat motion for centuries?

A physicist at the end of the last century would have been at a loss to answer this question, if he was prepared to draw only on those laws of Nature which he could explain and which he really understood. Perhaps, indeed, after a short reflection on the statistical situation he would have answered (correctly, as we shall see): These material structures can only be molecules. Of the existence, and sometimes very high stability, of these associations of atoms, chem-

让我再一次分享这个令人惊讶的事例。哈布斯堡王朝好几个皇室成员有着一种特征性的下唇畸形（“哈布斯堡之唇”）。维也纳的帝国学院在王室的赞助下，对这一遗传性状进行了全面细致的研究并发表了研究结果，还配上了旧时的肖像。事实证明，这一性状是嘴唇正常形态的真正的孟德尔“等位基因”。让我们把焦点集中在16世纪一个王族成员和他生活在19世纪的后裔的肖像上，我们完全有把握地假设，造成这种异常特征的基因结构物质已经在好几个世纪中代代相传，在两代之间为数不多的细胞分裂中稳定复制。而且，参与相关基因结构的原子数量很可能与X射线实验中的在相同的数量级。这一基因一直保存在98 ℉（36.5 ℃）的温度下。我们该如何理解它在几个世纪以来一直未受热运动的干扰而趋向无序呢？

上个世纪末的物理学家对这个问题的回答会感到困惑，如果他准备只用那些他能解释的、他真正理解的自然规律去回答这个问题。也许，的确，在对统计情况作一番不长时间的思考之后，他会这样回答（也正如我们会看到的那样，是正确的）：这些物质结构只能是分子。对于这些相关的原子集合体的存在，有时甚至是具有高度稳定性的，化学这一学科在当时已经积累

istry had already acquired a widespread knowledge at the time. But the knowledge was purely empirical. The nature of a molecule was not understood—the strong mutual bond of the atoms which keeps a molecule in shape was a complete conundrum（难题）to everybody. Actually, the answer proves to be correct. But it is of limited value as long as the enigmatic biological stability is traced back only to an equally enigmatic chemical stability. The evidence that two features, similar in appearance, are based on the same principle, is always precarious as long as the principle itself is unknown.

了广泛的知识，但这些了解还只是纯粹经验性的。还不能理解分子的性质 —— 维持分子结构的原子之间的强键当时对我们所有人来说都还是个谜。事实上，这个答案已被证明是正确的。但如果只是把生物稳定性之谜归因于同样神秘的化学稳定性，它的价值是有限的。将看起来表面相似的两种特性归于同样的原理基础，那么这些证据就是靠不住的，只要这一原理自身尚还未知。

EXPLICABLE BY QUANTUM THEORY

In this case it is supplied by quantum theory. In the light of present knowledge, the mechanism of heredity is closely related to, 48 nay, founded on, the very basis of quantum theory. This theory was discovered by Max Planck in 1900. Modern genetics can be dated from the rediscovery of Mendel's paper by de Vries, Correns and Tschermak (1900) and from de Vries's paper on mutations (1901—3). Thus the births of the two great theories nearly coincide, and it is small wonder that both of them had to reach a certain maturity before the connection could emerge. On the side of quantum theory it took more than a quarter of a century till in 1926—7 the quantum theory of the chemical bond was outlined in its general principles by W. Heitler and F. London. The Heitler-London theory involves the most subtle and intricate conceptions of the latest development of quantum theory (called 'quantum mechanics' or 'wave mechanics'). A presentation without the use of calculus is well-nigh impossible or would at least require another little volume like this. But fortunately, now that all work has been done and has served to clarify our thinking, it seems to be possible to point out in a more direct manner the connection between 'quantum jumps' and mutations, to pick out at the moment the most conspicuous item. That is what we attempt here.

量子理论的解释

这种情况下可借用量子理论。

根据现有的知识，遗传机制与量子理论息息相关，实实在在地建立在量子理论的基础之上。量子理论是普朗克在1900年建立的。现代遗传学也可上溯到1900年德弗里斯、科伦斯、切尔玛克对孟德尔论文的再发现，以及德弗里斯关于突变的论文（1901—1903）。可以看出，这两个伟大的学说，几乎是同时问世的。只有这两个理论都发展到一定程度时才会融为一体，这一点无须惊叹。量子理论这一方面的进展，差不多花了四分之一个世纪的时间，直到1926—1927年，化学键的量子理论基本原理才由海特勒和伦敦予以阐明。海特勒-伦敦理论涵盖了当时量子理论最新进展中最为精细并且最为复杂的概念——“量子力学”或“波动力学”。

要想阐明这个问题而不采用微积分是不可能的，或者至少需要另写一本这样的小册子。幸运的是，既然所有的工作都已做了，都可以用来阐明我们的想法，似乎有可能以更直接的方式阐明“量子跃迁”和突变之间的联系，并指出其亮点所在。这正是我们试图要做的事。

QUANTUM THEORY—DISCRETE STATES—QUANTUM JUMPS

The great revelation of quantum theory was that features of discreteness were discovered in the Book of Nature, in a context in which anything other than continuity seemed to be absurd according to the views held until then.

The first case of this kind concerned energy. A body on the large scale changes its energy continuously. A pendulum（钟摆）, for instance, that is set swinging is gradually slowed down by the resistance of the air. Strangely enough, it proves necessary to admit that a system of the order of the atomic scale behaves differently. On grounds upon which we cannot enter here, we have to assume that a small system can by its very nature possess only certain discrete amounts of energy, called its peculiar energy levels. The transition from one state to another is a rather mysterious event, which is usually called a 'quantum jump'.

49 But energy is not the only characteristic of a system. Take again our pendulum, but think of one that can perform different kinds of movement, a heavy ball suspended by a string from the ceiling. It can be made to swing in a north-south or east-west or any other direction or in a circle or in an ellipse. By gently blowing the ball with a bellows, it can be made to pass continuously from one state of motion to any other.

量子理论：离散态和量子跃迁

量子理论最主要的启示，是在“自然之书”中发现的离散态的特点，在这种情况下，根据当时的观点，除了连续性之外，其他都是荒谬可笑的。

第一个这样的实例与能量有关。一个物体在大尺度上可以连续改变能量。比如，一个钟摆，让它摆动之后，会由于空气的阻力而逐渐变慢。奇怪的是，事实证明一个原子级的系统行为可不是这样。因为某些原因我们无法在这里详细描述，设想一个微小的系统，根据它的自然特性，只具有某些离散的能量，并将其称为特定能级。能量从这一状态到另一状态的跃迁是非常神秘的事件，它通常被称为“量子跃迁”。

但是，能量并不是一个系统的唯一特征。再拿我们的钟摆为例，设想一个钟摆能做各种不同方式的运动。天花板上悬着一根绳子，绳上挂了一个重球，可以使它一下子从南到北，一下子从西往东，另一下子转至任一其他的方向，或者作圆形或椭圆形运动。用吹风箱轻轻地对着这个球吹一下，这个球就可以以连续渐变的方式从一种状态的运动变为另一种状态的运动。

For small-scale systems most of these or similar characteristics—we cannot enter into details—change discontinuously. They are 'quantized', just as the energy is.

The result is that a number of atomic nuclei, including their bodyguards of electrons, when they find themselves close to each other, forming 'a system', are unable by their very nature to adopt any arbitrary configuration we might think of. Their very nature leaves them only a very numerous but discrete series of 'states' to choose from.[1] We usually call them levels or energy levels, because the energy is a very relevant part of the characteristic. But it must be understood that the complete description includes much more than just the energy. It is virtually correct to think of a state as meaning a definite configuration of all the corpuscles.

The transition from one of these configurations to another is a quantum jump. If the second one has the greater energy ('is a higher level'), the system must be supplied from outside with at least the difference of the two energies to make the transition possible. To a lower level it can change spontaneously, spending the surplus of energy in radiation.

1. I am adopting the version which is usually given in popular treatment and which suffices for our present purpose. But I have the bad conscience of one who perpetuates a convenient error. The true story is much more complicated, inasmuch as it includes the occasional indeterminateness with regard to the state the system is in.

对小尺度的系统来说，大多数特性，或与之相似的特性——我们还是不多讲那些细节——只能不连续地改变。我们说这些系统被“量子化”了，就像能量那样。

这个结果说明，一定数目的原子核，还有它们的卫兵——电子，当它们发觉彼此靠近时，就形成了一个“系统”。原子核，就它们的本质而言，并不能如我们所想象的那样形成任意一种构型。它们真正的性质是让它们自己只能在一系列大量的离散“状态”中选择[1]。我们通常称它们为“级”或“能级”，因为能量是其特性最关键的部分。但是，一定要理解，要想完整地描述这些状态，就需要了解很多很多能量之外的东西。把一个状态想象成一个由所有“微粒”确定的构型，基本上是正确的。

量子跃迁就是从一种构型转换到另一种构型。如果第二种构型具有更大的能量（“高能级”），这个系统一定要从外界接受能量，至少是这两个能量的差值，从而使转换成为可能。转向低能级的变化可以是自发的，通过辐射来消耗多余能量就行了。

1. 在这里我采取的方式，是通常用于大众科普的方式，能够达到我们现在的目的就行了。但是，我又担心犯了一个常见的错误。实际的情况其实要复杂得多，因为系统状态中的偶发事件使之具有不确定性。

MOLECULES

Among the discrete set of states of a given selection of atoms there need not necessarily but there may be a lowest level, implying 50 a close approach of the nuclei to each other. Atoms in such a state form a molecule. The point to stress here is, that the molecule will of necessity have a certain stability; the configuration cannot change, unless at least the energy difference, necessary to 'lift' it to the next higher level, is supplied from outside. Hence this level difference, which is a well-defined quantity, determines quantitatively the degree of stability of the molecule. It will be observed how intimately this fact is linked with the very basis of quantum theory, viz. with the discreteness of the level scheme.

I must beg the reader to take it for granted that this order of ideas has been thoroughly checked by chemical facts; and that it has proved successful in explaining the basic fact of chemical valency and many details about the structure of molecules, their binding-energies, their stabilities at different temperatures, and so on. I am speaking of the Heitler–London theory, which, as I said, cannot be examined in detail here.

分子

在所选择的原子的离散态中，不一定但可能有一个最低态，原子核能相互靠拢。原子就在这样一种状态下形成一个分子。这里要强调的重点是，这个分子必须具有一定的稳定性；构型不能改变，除非外界提供必要的能量差，能使它就近“上升”到一个较高的能级。因此，这一能级差是能够明确定量的，也就定量地决定了分子的稳定程度。我们可以观察到，这一事实与量子理论的真正基础，即能级的不连续性，是如何紧密联系的。

我必须恳请读者注意并尽管放心，我的这些思路已被化学证据所验证，而且已成功地解释了化学价存在的基本事实以及有关分子结构、分子不同温度下的结合能和稳定性等诸多细节。我正在讲的都是海特勒-伦敦的理论，正如我前面所说，无法在这里详细讨论。

THEIR STABILITY DEPENDENT ON TEMPERATURE

We must content ourselves with examining the point which is of paramount interest for our biological question, namely, the stability of a molecule at different temperatures. Take our system of atoms at first to be actually in its state of lowest energy. The physicist would call it a molecule at the absolute zero of temperature. To lift it to the next higher state or level a definite supply of energy is required. The simplest way of trying to supply it is to 'heat up' your molecule. You bring it into an environment of higher temperature ('heat bath'), thus allowing other systems (atoms, molecules) to impinge（撞击）upon it. Considering the entire irregularity of heat motion, there is no sharp temperature limit at which the 'lift' will be brought about with certainty and immediately. Rather, at any temperature (different from absolute zero) there is a certain smaller or greater chance for the lift to occur, the chance increasing of course with the temperature of the heat bath. The best way to express this chance is to indicate the average time you will have to wait until the lift takes place, the 'time of expectation'.

51 From an investigation, due to M. Polanyi and E. Wigner,[1] the 'time of expectation' largely depends on the ratio of two energies, one being just the energy difference itself that is required to effect the lift (let us write W for it), the other one characterizing the intensity of the heat motion at the temperature in question (let us write T

---

1. *Zeitschrift für Physik,* Chemie (A), Haber-Band (1928), *p. 439.*

稳定性取决于温度

我们应该研究生物学问题中极其重要的一点，即，分子在不同温度下的稳定性。

让我们的原子系统一开始处于能量最低的状态，物理学家称它为在绝对零度条件下的分子。要将它提升到一个较高的状态或能级，肯定需要一定的能量。最简单的供能方法是，给分子“加热”。你将它置于温度较高的环境之中（“热浴”），以便让其他系统（原子、分子）来撞击它。考虑到热运动的完全无规律性，并不存在准确的温度界限，使得“能级升级”肯定且立即发生。反而是，在任一温度下（非绝对零度），都有要么较小要么较大的概率发生“升级”。当然，这种概率随着“热浴”温度升高而增高。描述这一概率最好的方式是用“期望时间”，它表示你必须等待“升级”发生的平均时间。

据波拉尼和维格纳[1]的一个研究可以看出，“期望时间”在很大程度上取决于两个能量之比，一个正是能级升级所需的能量差（让我们以 W 来表示），另一个是在特定的温度下表示热

1.《物理学杂志，化学(A)》，[*Zeitschrift für Physik*，Chemie (A)，Haber-Band]，第439页，1928年。

for the absolute temperature and kT for the characteristic energy).[1] It stands to reason that the chance for effecting the lift is smaller, and hence that the time of expectation is longer, the higher the lift itself compared with the average heat energy, that is to say, the greater the ratio $W : kT$. What is amazing is how enormously the time of expectation depends on comparatively small changes of the ratio $W : kT$. To give an example (following Delbrück): for W 30 times kT the time of expectation might be as short as $\frac{1}{10}$ s, but would rise to 16 months when W is 50 times kT, and to 30,000 years when W is 60 times kT!

1. k is a numerically known constant, called Boltzmann' s constant; $\frac{3}{2}kT$ is the average kinetic energy of a gas atom at temperature T.

运动的强度的能量（让我们以 T 表示绝对温度，kT 表示特征能量）[1]。按说，引起“升级”的概率越小，“期望时间”就越长，相对于平均热能的能级升级越高，也就是说，$W:kT$的比值就越大。令人惊奇的是，“期望时间”在很大程度上依赖于相对较小的$W:kT$ 比值的变化。再举一个例子（德尔布吕克用过的）：W是kT 的30倍时，期望时间则短至十分之一秒，若W是kT的50倍时，期望时间则变为16个月，而当W是kT的60倍时，期望时间将是3万年。

1. k是已知常数，称为玻尔兹曼常数；$(3/2)\,kT$ 是在温度为T时一个气体原子的平均动能。

MATHEMATICAL INTERLUDE

It might be as well to point out in mathematical language—for those readers to whom it appeals—the reason for this enormous sensitivity to changes in the level step or temperature, and to add a few physical remarks of a similar kind. The reason is that the time of expectation, call it t, depends on the ratio W/kT by an exponential function, thus

$$t = \tau e^{W/kT}$$

τ is a certain small constant of the order of 10^{-13} or 10^{-14} s. Now, this particular exponential function is not an accidental feature. It recurs again and again in the statistical theory of heat, forming, as it were, its backbone. It is a measure of the improbability of an energy amount as large as W gathering accidentally in some particular part of the system, and it is this improbability which increases so enormously when a considerable multiple of the 'average energy' kT is required.

52 Actually a $W = 30kT$ (see the example quoted above) is already extremely rare. That it does not yet lead to an enormously long time of expectation (only $\frac{1}{10}$ s. in our example) is, of course, due to the smallness of the factor τ. This factor has a physical meaning. It is of the order of the period of the vibrations which take place in the system all the time. You could, very broadly, describe this factor as meaning that the chance of accumulating the required amount W, though very small, recurs again and again 'at every vibration', that is to say, about 10^{13} or 10^{14} times during every second.

数学的插曲

我们也可以借助数学语言 —— 向那些对此颇感兴趣的读者 —— 解释这种对能级变化或温度变化极为敏感的原因，同时再补充几个类似的物理学解释。原因是，期望时间，称为t，取决于比值W/kT的指数函数形式：

$$t=\tau e^{W/kT}$$

这里τ是相当于10^{-13}或10^{-14}秒这样数量级的非常微小的常数。现在，这个特定的指数函数并非偶然。它在热统计理论中反复出现，可以说是构成热统计理论的支柱。它可以度量系统中的某个特定部分偶然聚集像W这么巨大的能量时的不可能性概率。当系统需要相当多倍数的“平均能”kT时，这个不可能性概率便极大地增加了。

实际上，$W = 30\ kT$（见前面引用的例子）已是极为罕见了。它还没有导致极长的期望时间（在我们的例子中只有1/10秒），当然，这是因为系数τ很小。这个因子具有物理学的意义，它代表整个时间内这个系统里发生的振动周期的数量级。你可以粗略地将其认定为积累所需W的概率。它虽然很小，但在“每一次振动中”都会重复出现，亦即每秒内振动大约10^{13}或10^{14}次。

FIRST AMENDMENT

In offering these considerations as a theory of the stability of the molecule it has been tacitly assumed that the quantum jump which we called the 'lift' leads, if not to a complete disintegration, at least to an essentially different configuration of the same atoms—an isomeric molecule, as the chemist would say, that is, a molecule composed of the same atoms in a different arrangement (in the application to biology it is going to represent a different 'allele' in the same 'locus' and the quantum jump will represent a mutation).

To allow of this interpretation two points must be amended in our story, which I purposely simplified to make it at all intelligible. From the way I told it, it might be imagined that only in its very lowest state does our group of atoms form what we call a molecule and that already the next higher state is 'something else'. That is not so. Actually the lowest level is followed by a crowded series of levels which do not involve any appreciable change in the configuration as a whole, but only correspond to those small vibrations among the atoms which we have mentioned above. They, too, are 'quantized', but with comparatively small steps from one level to the next. Hence the impacts of the particles of the 'heat bath' may suffice to set them up already at fairly low temperature. If the molecule is an extended structure, you may conceive these vibrations as high-frequency sound waves, crossing the molecule without doing it any harm.

第一项修正

在把这些考虑作为分子稳定性的理论时，我们已经默认了这样一个假设，即我们称之为“升级”的量子跃迁，即使不会引发分子完全解离，那么至少也会导致相同的原子形成性质上不同的构型。这种不同的构型，即化学家们所说的同分异构分子，由相同的原子按照不同的排列组成的分子（应用到生物学上，表示同一“位点”上的不同“等位基因”，量子跃迁则可用来表示突变）。

要使这一解释成立，我们的理论必须做两项修正，我有意将其简化，以使其完全可以理解。依前所述，我们也许可以这样设想，一组原子只有处在能量最低的状态时，才能形成我们所说的分子，而最邻近的高能量状态已经是“别的什么东西”了。事实并非如此。实际上，最低能级后面还有着一系列密集的能级，它们并不导致整个构型发生任何明显的变化，只不过引起我们前面叙述的原子间微弱的振动。这些能级也已“量子化”，只是相邻能级之间有小小的阶差而已。因此，“热浴”中粒子的作用可能足以使它们处于相当低的温度。如果分子具有一种延展结构，你可以把这些振动设想成高频声波，穿过分子而不对它造成任何伤害。

53 So the first amendment is not very serious: we have to disregard the 'vibrational fine-structure' of the level scheme. The term 'next higher level' has to be understood as meaning the next level that corresponds to a relevant change of configuration.

SECOND AMENDMENT

The second amendment is far more difficult to explain, because it is concerned with certain vital, but rather complicated, features of the scheme of relevantly different levels. The free passage between two of them may be obstructed, quite apart from the required energy supply; in fact, it may be obstructed even from the higher to the lower state.

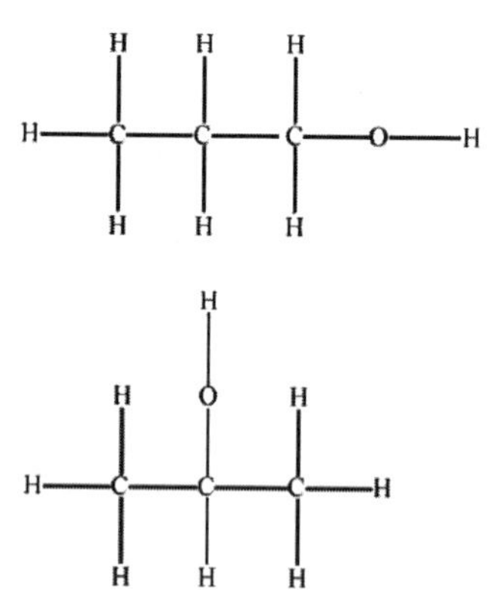

Fig. 11 . The two isomers of propyl-alcohol.

因此，第一项修正并不那么重要：我们不必理会能级图的“振动的精细结构”，而应把“下一个较高能级”理解为与构型改变相对应的下一个能级。

第二项修正

第二项修正的解释要困难得多，因为它关系到各个不同能级图的一些极为关键而又相当复杂的特性。两个能级之间的自由通路可能会受到损坏，与所需的能量毫不相关；事实上，从高能级跃迁到低能级的时候也可能会受到阻碍。

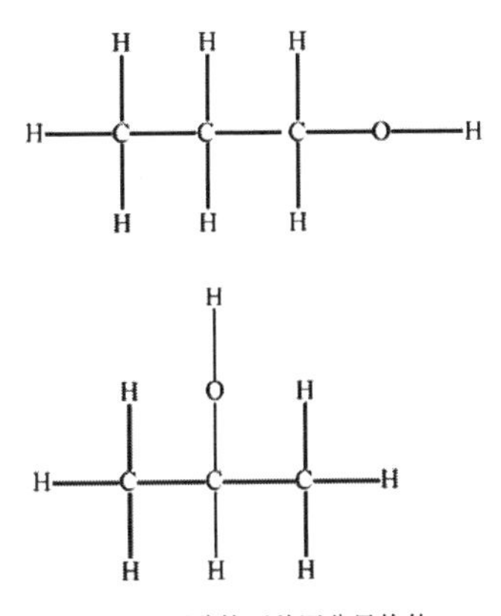

图11　丙醇的两种同分异构体

Let us start from the empirical facts. It is known to the chemist that the same group of atoms can unite in more than one way to form a molecule. Such molecules are called isomeric ('consisting of the same parts', ἴσος=same, μέρος =part). Isomerism is not an 54 exception, it is the rule. The larger the molecule, the more isomeric alternatives are offered. Fig. 11 shows one of the simplest cases, the two kinds of propylalcohol, both consisting of 3 carbons (C), 8 hydrogens (H), 1 oxygen (O).[1] The latter can be interposed between any hydrogen and its carbon, but only the two cases shown in our figure are different substances. And they really are. All their physical and chemical constants are distinctly different. Also their energies are different, they represent 'different levels'.

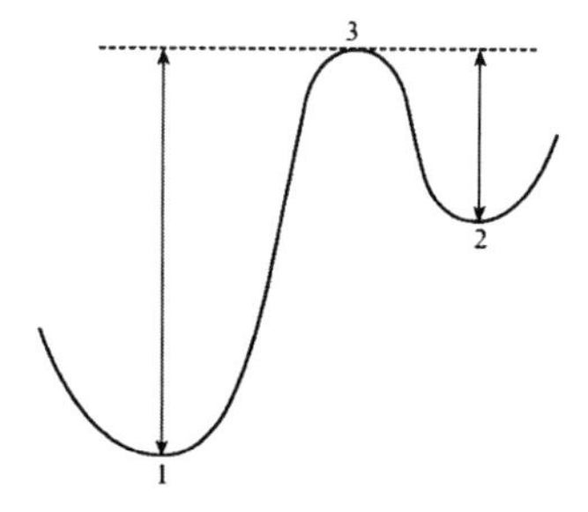

Fig. 12. Energy threshold (3) between the isomeric levels (1) and (2). The arrows indicate the minimum energies required for transition.

1. Models, in which C, H and O were represented by black, white and red wooden balls respectively, were exhibited at the lecture. I have not reproduced them here, because their likeness to the actual molecules is not appreciably greater than that of Fig. 11.

让我们从经验性的事实开始。化学家都知道，同一组原子可以通过多种方式结合形成分子。这些分子被称为同分异构体（isomeric,“由相同的部件构成”；ίσος=“相同的”，μέρος =“部件”）。同分异构不是一种例外，而是一种规律。分子越大，存在的同分异构体就越多。图11是其中一种最简单的情况，即两种丙醇，都是由3个碳原子（C）、8个氢原子（H）和1个氧原子（O）构成[1]。后者（氧原子）可以介于任何氢原子和碳原子之间，但只有像我们图中的两个物质，才是真正不同的物质。它们所有的物理常数和化学常数都存在显著差异。它们的能量也不同，代表了“不同的能级”。

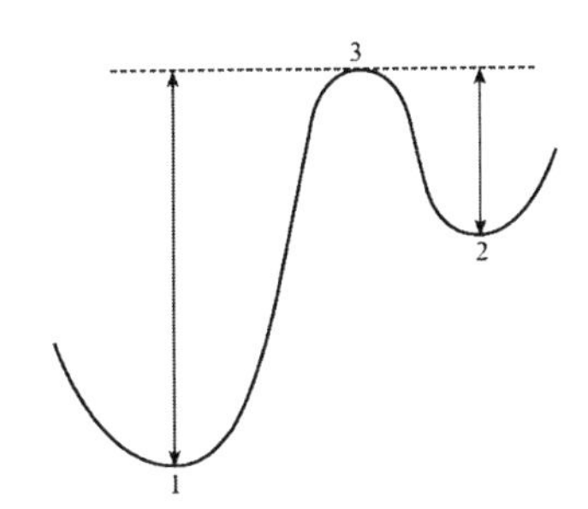

图12　同分异构体能级（1）和（2）之间的能量阈值（3）
箭头表示发生跃迁所需的最低能量

1. 那天在演讲时我还展示了相关的分子模型，其中C、H、O原子分别用黑色、白色和红色的木头圆球表示。这里我就不再演示了，因为它们和实际分子的相似性比图11也好不了多少。

The remarkable fact is that both molecules are perfectly stable, both behave as though they were 'lowest states'. There are no spontaneous transitions from either state towards the other.

The reason is that the two configurations are not neighbouring configurations. The transition from one to the other can only take place over intermediate configurations which have a greater energy than either of them. To put it crudely, the oxygen has to be extracted from one position and has to be inserted into the other. There does not seem to be a way of doing that without passing through configurations of considerably higher energy. The state of affairs is some-
55 times figuratively pictured as in Fig. 12, in which 1 and 2 represent the two isomers, 3 the 'threshold' between them, and the two arrows indicate the 'lifts', that is to say, the energy supplies required to produce the transition from state 1 to state 2 or from state 2 to state 1, respectively.

Now we can give our 'second amendment', which is that transitions of this 'isomeric' kind are the only ones in which we shall be interested in our biological application. It was these we had in mind when explaining 'stability' on pp. 49–51. The 'quantum jump' which we mean is the transition from one relatively stable molecular configuration to another. The energy supply required for the transition (the quantity denoted by W) is not the actual level difference, but the step from the initial level up to the threshold (see the arrows in Fig. 12).

值得关注的事实是，这两个分子都极其稳定，似乎都表现为“最低的能级状态”。不存在从任何一种状态到另一种状态的自发转换。

其原因在于，这两种构型并不是相邻能级的构型。要从一种构型转换为另一种构型，必须经过若干中间构型，而这些中间构型的能量要高于两者中的任何一种。粗略地说，必须把氧原子从一个位置抽出来，插到另一个位置上。似乎没有一种途径可以在不通过相当高能量的构型的情况下做到这一点。这种状态有时可以用图12来显示，其中1和2代表两个同分异构体，3代表它们之间的“阈值”，两个箭头表示“跃迁”量，即产生从状态1到状态2或者从状态2到状态1的转换中分别所需的能量。

现在可以给出我们的“第二项修正”了，那就是只有这种“同分异构体”的转换是我们在生物学应用中最感兴趣的。这就是我们在本章第4-5节（原文第49-51页）中解释“稳定性”时想到的。所谓的“量子跃迁”，我们指的是从一种相对稳定的分子构型转变为另一种相对稳定的分子构型。发生转变所需的能量（用W表示）并不是真正的能级差，而是从初始能级上升到阈值的能级差（见图12中的箭头）。

Transitions with no threshold interposed between the initial and the final state are entirely uninteresting, and that not only in our biological application. They have actually nothing to contribute to the chemical stability of the molecule. Why? They have no lasting effect, they remain unnoticed. For, when they occur, they are almost immediately followed by a relapse（退回原状）into the initial state, since nothing prevents their return.

在初态和终态之间没有阈值介入的转换是毫无意义的，这不仅体现在我们的生物学应用中，它们实际上对分子的化学稳定性也没有任何贡献。为什么呢？因为它们不会产生任何持久的影响，难以引起人们的注意。当它们发生时，几乎是立刻又倒退到了初始状态，因为没有什么能够阻碍它们恢复原态。

CHAPTER 5
Delbrück's Model Discussed and Tested

Sane sicut lux seipsam et tenebras manifestat, sic veritas norma sui et falsi est.[1]

SPINOZA, *Ethics*, Pt II, Prop. 43.

THE GENERAL PICTURE OF THE HEREDITARY SUBSTANCE

56 From these facts emerges a very simple answer to our question, namely: Are these structures, composed of comparatively few atoms, capable of withstanding for long periods the disturbing influence of heat motion to which the hereditary substance is continually exposed? We shall assume the structure of a gene to be that of a huge molecule, capable only of discontinuous change, which consists in a rearrangement of the atoms and leads to an isomeric[2] molecule. The rearrangement may affect only a small region of the gene, and a vast number of different rearrangements may be possible. The energy thresholds, separating the actual configuration from any possible

1. Truly, as light manifests itself and darkness, thus truth is the standard of itself and of error.
2. For convenience I shall continue to call it an isomeric transition, though it would be absurd to exclude the possibility of any exchange with the environment.

第五章
德尔布吕克模型的讨论和验证

光线展示了自己，也暴露了黑暗，

真理证明了自己，也区别了谬误。

——斯宾诺莎《伦理学》第二卷，命题43

遗传物质的概貌

从前面例举的事实，可以直观且粗略地回答我们的问题，那就是：这些遗传结构只是由数目有限的原子组成，能够长时间地承受热运动对遗传物质的影响吗？

我们应该这样设想，基因结构本身就是一个巨大的分子，能够进行不连续的改变，即原子的重排，并组成一个同分异构的分子[1]。这一重排也许只影响这个基因的一个小小的区域，

1. 为了方便起见，我将继续称它为同分异构转换，尽管这里不考虑它和环境相互转换的可能性是荒谬的。

isomeric ones, have to be high enough (compared with the average heat energy of an atom) to make the change-over a rare event. These rare events we shall identify with spontaneous mutations.

The later parts of this chapter will be devoted to putting this general picture of a gene and of mutation (due mainly to the German 57 physicist M. Delbrück) to the test, by comparing it in detail with genetical facts. Before doing so, we may fittingly make some comment on the foundation and general nature of the theory.

而且可能发生了大量不同的重排。将实际的构型从任何可能的同分异构分子中区别开来的能量阈值一定要足够大（与一个原子的平均热能相比），大到足够将之转化为一个罕见事件，这些罕见事件就是我们所说的自发突变。

本章的后面部分将致力于对基因和突变的概貌（主要是德国物理学家德尔布吕克的模型）进行验证，通过将其与遗传学证据进行详细比较。在此之前，我们还是先对该理论的基础和一般性质作一些说明。

THE UNIQUENESS OF THE PICTURE

Was it absolutely essential for the biological question to dig up the deepest roots and found the picture on quantum mechanics? The conjecture that a gene is a molecule is today, I dare say, a commonplace. Few biologists, whether familiar with quantum theory or not, would disagree with it. On p. 47 we ventured to put it into the mouth of a pre-quantum physicist, as the only reasonable explanation of the observed permanence. The subsequent considerations about isomerism, threshold energy, the paramount role of the ratio $W : kT$ in determining the probability of an isomeric transition—all that could very well be introduced on a purely empirical basis, at any rate without drawing on quantum theory. Why did I so strongly insist on the quantum-mechanical point of view, though I could not really make it clear in this little book and may well have bored many a reader?

Quantum mechanics is the first theoretical aspect which accounts from first principles for all kinds of aggregates of atoms actually encountered in Nature. The Heitler–London bondage is a unique, singular feature of the theory, not invented for the purpose of explaining the chemical bond. It comes in quite by itself, in a highly interesting and puzzling manner, being forced upon us by entirely different considerations. It proves to correspond exactly with the observed chemical facts, and, as I said, it is a unique feature, well enough understood to tell with reasonable certainty that 'such a

基因的独特性

对于生物学问题来说，挖掘最深层的根源并在量子力学这里找到答案是绝对必要的吗？

今天，基因是分子的推测，我敢说，这已成为共识。几乎没有生物学家，不论其是否熟悉量子理论，不赞同这一点。

在第四章的第2节（原文第47页），我们大胆地借前量子物理学家之口，来说明只有这个理论才能合理地解释所观察到的基因持久性。接着我们开始讨论同分异构体和能量阈值，$W:kT$ 在确定同分异构转换概率方面的极其重要作用等——这些本来都可以很好地在纯粹的经验基础之上引入，而不需要利用任何量子理论。为什么我还要如此强烈地坚持量子力学的观点呢？尽管我在这本小册子中无法真正讲明白量子力学，而且还可能使许多读者感到厌烦。

量子力学是第一个从基本原理出发，解释自然界中实际遇到的各种原子聚集的理论。海特勒-伦敦键是该理论唯一且独有的特征，它不是为了解释化学键而提出来的。它以一种非常有趣而又匪夷所思的方式自发出现，出于五花八门的考虑强加给我们。但是，现已证明，它与观察到的化学论据精确吻合，而

thing could not happen again' in the further development of quantum theory.

Consequently, we may safely assert that there is no alternative to the molecular explanation of the hereditary substance. The physical aspect leaves no other possibility to account for its permanence. If the Delbrück picture should fail, we would have to give up further attempts. That is the first point I wish to make.

且正如我所说，它是独有的，能被充分理解，可以相当肯定地说，在量子理论的进一步发展中，“这样的事不可能再次发生”。

因此，我们很有把握地断言，除了将遗传物质解释为分子以外，没有什么比它更合适的了。物理方面没有留下其他的可能性来解释它的持久性。倘若德尔布吕克的理论不成立，我们将不得不放弃进一步的尝试。这就是我想说的第一点。

SOME TRADITIONAL MISCONCEPTIONS

58 But it may be asked: Are there really no other endurable structures composed of atoms except molecules? Does not a gold coin, for example, buried in a tomb for a couple of thousand years, preserve the traits of the portrait stamped on it? It is true that the coin consists of an enormous number of atoms, but surely we are in this case not inclined to attribute the mere preservation of shape to the statistics of large numbers. The same remark applies to a neatly developed batch of crystals we find embedded in a rock, where it must have been for geological periods without changing.

That leads us to the second point I want to elucidate. The cases of a molecule, a solid, a crystal are not really different. In the light of present knowledge they are virtually the same. Unfortunately, school teaching keeps up certain traditional views, which have been out of date for many years and which obscure the understanding of the actual state of affairs.

Indeed, what we have learnt at school about molecules does not give the idea that they are more closely akin to the solid state than to the liquid or gaseous state. On the contrary, we have been taught to distinguish carefully between a physical change, such as melting or evaporation in which the molecules are preserved (so that, for example, alcohol, whether solid, liquid or a gas, always consists of the same molecules, C_2H_6O), and a chemical change, as, for example, the burning of alcohol,

几个错误的传统观念

但也许有人会问：除了分子以外，真的就没有其他由原子组成的并且可以长时间保持稳定的结构吗？例如，一枚金币，埋在坟墓中几千年了，印在上面的肖像不是也逼真地保留下来了吗？诚然，金币也是由无数原子组成的。但是在这个例子里，我们显然不会把金币形状的持久归因于大量的统计。这种说法也适用于我们发现的那些嵌在岩石中的长年累日形成的整整齐齐的晶体，它历经数个地质时代而没有改变。

这就引出了我要说明的第二点。实际上，一个分子、一个固体和一块晶体的情况并没有实质性的差别。从现代知识来看，它们实质上也没有什么不同。遗憾的是，学校的教学中还保持着好多年前已过时了的传统观念，从而混淆了人们对真实情况的理解。

的确，从学校里学到的关于分子的知识并没有告诉我们，分子更接近固态而不是液态或气态。与之相反，学校教导我们要仔细区分物理变化（例如熔化或蒸发。分子在这些变化中是保持不变的。因此，比如酒精，不管它是固体、液体还是气体，总是由相同的分子C_2H_6O组成）和化学变化（如酒精的燃烧）：

$$C_2H_6O + 3O_2 = 2CO_2 + 3H_2O,$$

where an alcohol molecule and three oxygen molecules undergo a rearrangement to form two molecules of carbon dioxide and three molecules of water.

About crystals, we have been taught that they form three-fold periodic lattices, in which the structure of the single molecule is sometimes recognizable, as in the case of alcohol and most organic compounds, while in other crystals, e.g. rock-salt (NaCI), NaCI mol-

59 ecules cannot be unequivocally（明确地）delimited, because every Na atom is symmetrically surrounded by six CI atoms, and vice versa, so that it is largely arbitrary what pairs, if any, are regarded as molecular partners.

Finally, we have been told that a solid can be crystalline or not, and in the latter case we call it amorphous.

$$C_2H_6O + 3O_2 = 2CO_2 + 3H_2O,$$

在这里，1个酒精分子同3个氧分子作用，经过原子重排而生成了2个二氧化碳分子和3个水分子。

关于晶体，我们学到的是它们会形成三维的周期性晶格，其中单个分子的结构有时是可识别的，比如乙醇和大多数有机化合物；而在其他晶体中，例如岩盐（氯化钠）的晶格中就无法明确地区别出单个的氯化钠分子。因为每一个钠原子都被6个氯原子对称地包围，反之亦然。所以，到底将哪一对氯原子和钠原子视作一个分子是相当随意的。

最后，我们还学到，一个固体可以是晶体，也可以不是晶体，后者我们称其为非晶态。

DIFFERENT 'STATES' OF MATTER

Now I would not go so far as to say that all these statements and distinctions are quite wrong. For practical purposes they are sometimes useful. But in the true aspect of the structure of matter the limits must be drawn in an entirely different way. The fundamental distinction is between the two lines of the following scheme of 'equations':

molecule = solid = crystal.

gas = liquid = amorphous.

We must explain these statements briefly. The so-called amorphous solids are either not really amorphous or not really solid. In 'amorphous' charcoal fibre the rudimentary（基本的）structure of the graphite（石墨）crystal has been disclosed by X-rays. So charcoal is a solid, but also crystalline. Where we find no crystalline structure we have to regard the thing as a liquid with very high 'viscosity' (internal friction). Such a substance discloses by the absence of a well-defined melting temperature and of a latent heat of melting that it is not a true solid. When heated it softens gradually and eventually liquefies without discontinuity. (I remember that at the end of the first Great War we were given in Vienna an asphalt-like（沥青）substance as a substitute for coffee. It was so hard that one had to use a chisel or a hatchet to break the little brick into pieces, when it would show a smooth, shell-like cleavage. Yet, given time, it would behave as a liquid, closely packing the lower part of a vessel in which you

物质的不同“状态”

现在，我还不打算说所有的这些说法和区分都是错误的。就实用目的来说，它们有时候还是有用的。但是，就物质的真实结构而言，这些界限必须以完全不同的方式加以界定。最基本的区分在于下面的两行“等式”：

分子=固体=晶体
气体=液体=非晶态的。

我们有必要简单地解释一下这些说法。所谓的非晶态固体，要么不是真正的非晶态，要么就不是真正的固体。在“非晶态的”木炭纤维中，石墨晶体的基本结构已用X射线来揭示。所以，木炭既是固体也是晶体。对于还没有发现晶体结构的物质，就必须把它看成一种高“黏性”（内摩擦）的液体。此类物质没有明确的熔化温度，也没有熔化的潜热，所以并非真正的固体。当它被加热时，会逐渐软化并最终液化，没有间断（我还记得在第一次世界大战结束时，我们在维也纳得到了一种有点像沥青的东西，作为咖啡的替代物。它是如此坚硬，以至于人们不得不用凿子或小斧子将小砖头一样的“咖啡”打成碎片，这时它将显示光滑的贝壳状裂缝。但是过了一段时间后，它就又像液体那样

were unwise enough to leave it for a couple of days.)

The continuity of the gaseous and liquid state is a well-known
60 story. You can liquefy any gas without discontinuity by taking your
way 'around' the so-called critical point. But we shall not enter on
this here.

牢牢地粘在容器的底部，而你却不知底细地把它搁上几天）。

气态和液态的连续性是众所周知的。你可以“绕过”所谓的临界点，使任何气体没有间断地液化。在这里就不多说了。

THE DISTINCTION THAT REALLY MATTERS

We have thus justified everything in the above scheme, except the main point, namely, that we wish a molecule to be regarded as a solid = crystal.

The reason for this is that the atoms forming a molecule, whether there be few or many of them, are united by forces of exactly the same nature as the numerous atoms which build up a true solid, a crystal. The molecule presents the same solidity of structure as a crystal. Remember that it is precisely this solidity on which we draw to account for the permanence of the gene!

The distinction that is really important in the structure of matter is whether atoms are bound together by those 'solidifying'Heitler-London forces or whether they are not. In a solid and in a molecule they all are. In a gas of single atoms (as e.g. mercury vapour) they are not. In a gas composed of molecules, only the atoms within every molecule are linked in this way.

THE APERIODIC SOLID

A small molecule might be called 'the germ of a solid'. Starting from such a small solid germ, there seem to be two different ways of building up larger and larger associations. One is the comparatively dull way of repeating the same structure in three directions again and again. That is the way followed in a growing crystal. Once the

真正重要的区别

到这里，我们的确已经阐述了前面所说的一切，除了这一要点，我们希望把一个分子看成是一个固体晶体，也就是固体=晶体。

其理由是，组成一个分子的原子，不管数目多少，都是由一种性质完全相同的力结合在一起的，就像组成真正的固体、晶体的众多原子一样。这种分子具有与晶体相同的稳固结构。请记住，我们正是基于稳固性来解释基因的持久性！

细小物质结构中真正重要的区别在于原子是否被那些起稳固作用的海特勒-伦敦力结合在一起。在固体和分子中，原子都是这样结合的。在单原子气体（如汞蒸气）中就不是这样了。在分子组成的气体中，只有每个分子中的原子是以这种方式连在一起的。

非周期性固体

一个小分子也许可以称为“固体的胚芽”。从这样一个小的固体胚芽开始，似乎有两种不同方式来形成越来越大的聚合体。第一种方式是相对单调的方式，在三维方向不断重复同样的结构。这就是生长中的晶体遵循的方式。一旦周期性建立起来，聚

periodicity is established, there is no definite limit to the size of the aggregate. The other way is that of building up a more and more extended aggregate without the dull device of repetition. That is the case of the more and more complicated organic molecule in which every atom, and every group of atoms, plays an individual role, not entirely equivalent to that of many others (as is the case in a periodic 61 structure). We might quite properly call that an aperiodic crystal or solid and express our hypothesis by saying: We believe a gene—or perhaps the whole chromosome fibre[1] —to be an aperiodic solid.

1. That it is highly flexible is no objection; so is a thin copper wire.

合体的大小就没有明确的限度。另一种方式是建立越来越大的聚合体，而不是单一的重复。复杂多变的有机分子就是如此，其中的每一个原子、每一个原子团都起着各自的作用，和其他分子并不完全一样（如同一个周期性结构那样）。我们可以恰如其分地称之为非周期性晶体或固体，而且我们的假设就可以表述为：我们相信一个基因——或许整个染色体纤丝[1]——就是一个非周期性的固体。

1. 毫无疑问染色体纤丝非常柔韧，就像一根纤细的铜丝那样。

THE VARIETY OF CONTENTS COMPRESSED IN THE MINIATURE CODE

It has often been asked how this tiny speck of material, nucleus of the fertilized egg, could contain an elaborate code-script involving all the future development of the organism. A well-ordered association of atoms, endowed with sufficient resistivity to keep its order permanently, appears to be the only conceivable material structure that offers a variety of possible ('isomeric') arrangements, sufficiently large to embody a complicated system of 'determinations' within a small spatial boundary. Indeed, the number of atoms in such a structure need not be very large to produce an almost unlimited number of possible arrangements. For illustration, think of the Morse code. The two different signs of dot and dash in well-ordered groups of not more than four allow of thirty different specifications. Now, if you allowed yourself the use of a third sign, in addition to dot and dash, and used groups of not more than ten, you could form 88,572 different 'letters'; with five signs and groups up to 25, the number is 372,529,029,846,191,405.

It may be objected that the simile is deficient, because our Morse signs may have different composition (e.g. ·−−− and ··−−) and thus they are a bad analogue for isomerism. To remedy this defect, let us pick, from the third example, only the combinations of exactly 25 symbols and only those containing exactly 5 out of each of the supposed 5 types (5 dots, 5 dashes, etc.). A rough count gives you

微型密码含义无穷

人们经常会问，一个受精卵的细胞核那么微小，怎么能够容纳一个那么精巧的、关系到有机体未来发育的所有指令的密码本。一个高度有序的原子集合体，被赋予足够的抵抗力来长久维持秩序的稳定性，这似乎是唯一可以接受的物质结构，它提供了各种可能（“同分异构的”）的排列，足以在一个很小的空间范围内容纳一个复杂的“决定性”系统。其实，这种结构无须那么多的原子，便能产生几乎无限的可能排列。

为了说明问题，试想一下摩尔斯电码。这种电码用点（“·”）和横（“–”）两个不同的符号，如果每一个有序的组合用的符号不超过4位，就可以排列组合成30种不同含义的电码。现在，除了点和横以外，如果你允许自己使用第三个符号，每一个组合用的字符不超过10个，就可以排列组合成88,572个不同的“字母”；如果用5个符号，每一个组合用的符号增加到25个，那么这个数字会是372,529,029,846,191,405。

可能有人会反驳，光我们的两个摩尔斯符号就可以有不同的组合（如 ·-- 和 ··-），因此与同分异构体作这样的类比并不恰当。为了弥补这个不足，让我们从第三个例子中挑选正好25个符号的组合，而且只能挑出5种符号，每个符号5种

the number of combinations as 62,330,000,000,000, where the zeros on the right stand for figures which I have not taken the trouble to compute.

Of course, in the actual case, by no means 'every' arrangement of the group of atoms will represent a possible molecule; moreover, 62 it is not a question of a code to be adopted arbitrarily, for the code-script must itself be the operative factor bringing about the development. But, on the other hand, the number chosen in the example (25) is still very small, and we have envisaged only the simple arrangements in one line. What we wish to illustrate is simply that with the molecular picture of the gene it is no longer inconceivable that the miniature code should precisely correspond with a highly complicated and specified plan of development and should somehow contain the means to put it into operation.

形式（5个点，5个横等）。粗略地算一下，字符串的组合是62,330,000,000,000个，后面的几个零代表的数字，我就不必费心去细算了。

当然，在实际情况中，一团原子的“每一”排列绝不能都代表一种可能的分子；而且，这也不是随便什么密码都能被采用的问题，因为密码本其本身还必定是发育过程中的操纵因子。但是，另一方面，上述例子中选用的数目（25个）还是太小了，而且我们只不过设想了一条线性的简单排列。我们想要说明的只是，有了基因的分子图，微型密码应该与特定的高度复杂的发育蓝图计划精确对应，并且应该以某种方式包含将其付诸实施的手段，这已经不再是不可想象的了。

COMPARISON WITH FACTS: DEGREE OF STABILITY; DISCONTINUITY OF MUTATIONS

Now let us at last proceed to compare the theoretical picture with the biological facts. The first question obviously is, whether it can really account for the high degree of permanence we observe. Are threshold values of the required amount—high multiples of the average heat energy kT—reasonable, are they within the range known from ordinary chemistry? That question is trivial; it can be answered in the affirmative without inspecting tables. The molecules of any substance which the chemist is able to isolate at a given temperature must at that temperature have a lifetime of at least minutes. (That is putting it mildly; as a rule they have much more.) Thus the threshold values the chemist encounters are of necessity precisely of the order of magnitude required to account for practically any degree of permanence the biologist may encounter; for we recall from p. 51 that thresholds varying within a range of about 1 : 2 will account for lifetimes ranging from a fraction of a second to tens of thousands of years.

But let me mention figures, for future reference. The ratios W/kT mentioned by way of example on p. 51, viz.

$$\frac{W}{kT}=30, 50, 60,$$

producing lifetimes of

$$\frac{1}{10}\text{s, 16 months, 30,000 years,}$$

稳定程度和突变的不连续性

现在，让我们最后把理论描述与生物学事实加以比较。第一个问题显然是，它能否真正解释我们观察到的高度持久性。所需要的阈值——平均热能kT的好几倍——是合理的吗？它们是否在普通化学知识的已知范畴之内？

这个问题是不难回答的，无须查表就可以给予肯定的解答。化学家能够在给定温度下分离出任何物质的分子，在该温度下分子至少有几分钟的存活寿命。（这是比较保守的说法，一般说来，它们的寿命要比这长得多。）所以，化学家获得的阈值必然恰好是解释生物学家所遇到的任何持久性的数量级；因为从第四章第5节（原文第51页）我们得知，在大约1∶2的范围内变化的阈值，可以解释从几分之一秒到数万年不等的寿命。

让我提供一些数字作为后面的参考。在第四章的例子中提到的W/kT比值，即

$$\frac{W}{kT}=30, 50, 60,$$

对应的寿命分别是

$$\frac{1}{10}\text{秒，16个月，30,000年，}$$

63 respectively, correspond at room temperature with threshold values of

0.9, 1.5, 1.8 electron-volts.

We must explain the unit 'electron-volt', which is rather convenient for the physicist, because it can be visualized. For example, the third number (1.8) means that an electron, accelerated by a voltage of about 2 volts, would have acquired just sufficient energy to effect the transition by impact. (For comparison, the battery of an ordinary pocket flash-light has 3 volts.)

These considerations make it conceivable that an isomeric change of configuration in some part of our molecule, produced by a chance fluctuation of the vibrational energy, can actually be a sufficiently rare event to be interpreted as a spontaneous mutation. Thus we account, by the very principles of quantum mechanics, for the most amazing fact about mutations, the fact by which they first attracted de Vries's attention, namely, that they are 'jumping' variations, no intermediate forms occurring.

分别地，在室温下对应的阈值为

0.9，1.5，1.8 电子伏特。

我们必须解释一下“电子伏特”这个单位。这对于物理学家是很方便的，因为它可以看得到。比如，上面第三个数字（1.8）意味着一个电子，被大约2伏特的电压加速，将获得足够的能量通过撞击产生跃迁（作为比较，请注意普通袖珍手电筒电池的电压为3伏）。

我们可以考虑接受这样的观点，在分子的某些部位，由振动能的偶然波动引起构型发生同分异构变化，实际上是一种极其罕见的事件，因而可以被解读为一次自发的突变。这样，从量子力学的基本原理出发，我们解释了关于突变的最令人惊奇的事实。这个事实最早吸引了德弗里斯的注意，即它们是“跳跃的”变异，没有中间形式发生。

STABILITY OF NATURALLY SELECTED GENES

Having discovered the increase of the natural mutation rate by any kind of ionizing rays, one might think of attributing the natural rate to the radio-activity of the soil and air and to cosmic radiation. But a quantitative comparison with the X-ray results shows that the 'natural radiation' is much too weak and could account only for a small fraction of the natural rate.

Granted that we have to account for the rare natural mutations by chance fluctuations of the heat motion, we must not be very much astonished that Nature has succeeded in making such a subtle choice of threshold values as is necessary to make mutation rare. For we have, earlier in these lectures, arrived at the conclusion that frequent mutations are detrimental to evolution. Individuals which, by mutation, acquire a gene configuration of insufficient stability, will have 64 little chance of seeing their 'ultra-radical', rapidly mutating descendancy survive long. The species will be freed of them and will thus collect stable genes by natural selection.

THE SOMETIMES LOWER STABILITY OF MUTANTS

But, of course, as regards the mutants which occur in our breeding experiments and which we select, *qua* mutants, for studying their offspring, there is no reason to expect that they should all show that very high stability. For they have not yet been 'tried out'—or, if they have, they have been 'rejected' in the wild breeds—possibly for too high mutability. At any rate, we are not at all astonished to learn that actually some of these mutants do show a much higher mutability than the normal 'wild' genes.

自然选择的基因及其稳定性

在发现了任何一种电离射线都会引起自然突变率升高的现象以后，人们也许会想到将自然突变率归咎于土壤和空气中的放射性以及宇宙射线。然而，与X射线的结果作定量比较却发现，“天然辐射”太弱了，只能解释自然突变率中很小的一部分。

倘若我们不得不解释罕见的自然突变是由热运动的偶然波动引起的，那么，对于自然界已成功做出的这样一个微妙的阈值选择，使突变恰好成为罕见现象，我们就不必大惊小怪。在前面的章节中，我们已得出这样的结论，频繁的突变对演化是有害的。个体由于突变获得了不够稳定的基因构型，那么，其遭受“超级辐射”、快速突变的后代几乎没有机会获得长寿。该物种将会淘汰这些个体，并通过自然选择把稳定的基因累积起来。

突变体并不都是那么稳定

但是，当然，对于在我们的育种实验中出现的突变体以及我们选择用来研究它们的后代的突变体，我们没有理由期望它们全都表现出很高的稳定性。因为它们还没有经受住“考验”—— 或者虽然经受住了“考验”，但它们在野生种中被“剔除”了 —— 可能是因为变异太大。无论如何，如果当真看到一些这样的突变体的突变性确实比正常的“野生型”基因高得多时，我们完全不必感到惊讶。

TEMPERATURE INFLUENCES UNSTABLE GENES LESS THAN STABLE ONES

This enables us to test our mutability formula, which was

$$t = \tau e^{W/kT}.$$

(It will be remembered that t is the time of expectation for a mutation with threshold energy W.) We ask: How does t change with the temperature? We easily find from the preceding formula in good approximation the ratio of the value of t at temperature $T + 10$ to that at temperature T

$$\frac{t_{T+10}}{t_T} = e^{-10W/kT^2}.$$

The exponent being now negative, the ratio is, naturally, smaller than 1. The time of expectation is diminished by raising the temperature, the mutability is increased. Now that can be tested and has been tested with the fly *Drosophila* in the range of temperature which the insects will stand. The result was, at first sight, surprising. The *low* mutability of wild genes was distinctly increased, but the comparatively *high* mutability occurring with some of the already 65 mutated genes was not, or at any rate was much less, increased. That is just what we expect on comparing our two formulae. A large value of W/kT, which according to the first formula is required to make t large (stable gene), will, according to the second one, make for a small value of the ratio computed there, that is to say for a considerable increase of mutability with temperature. (The actual values of

温度和基因的稳定性

这使我们能够检验突变性的公式，如下：

$$t=\tau e^{W/kT}。$$

（记住，t是一个阈值为W的突变的期望时间。）我们的疑问是：t是如何随温度而变化的？据上面的公式，我们可以很容易地得出，温度分别为$T+10$和T时，两者t值之比的最佳近似值为：

$$\frac{t_{T+10}}{t_T}=e^{-10W/kT^2}。$$

如果现在指数为负值，比值自然小于1。随着温度的升高，期望时间很快变短，可突变性却增加了。这是可以检测到的，而且已经在果蝇身上检测过了，在昆虫能承受的温度范围之内。乍一看，这一结果令人意外。原本突变性较低的野生型基因的突变性明显提高了，而此前就已出现过某些突变的基因，其原本较高的突变性却没有增加，或者不管怎么说提高率都非常低。

这正是我们比较这两个公式时所预期的结果。根据第一个公式，要想使t变大（稳定的基因），就要求W/kT的值增大；而根据第二个公式，W/kT的值增大了，则会使算出来的比值变小，就是说，突变可能性将随着温度增加而有较大的提高，这就解释了为何野生型基因的突变性随温度上升提高得很明显（实际

the ratio seem to lie between about $\frac{1}{2}$ and $\frac{1}{5}$. The reciprocal, 2.5, is what in an ordinary chemical reaction we call the van't Hoff factor.)

的比值大约在 $\frac{1}{2}$ 到 $\frac{1}{5}$ 之间，其倒数为2.5，后者就是普通化学反应中所说的范托夫因子）。

HOW X-RAYS PRODUCE MUTATION

Turning now to the X-ray-induced mutation rate, we have already inferred from the breeding experiments, first (from the proportionality of mutation rate, and dosage), that some single event produces the mutation; secondly (from quantitative results and from the fact that the mutation rate is determined by the integrated ionization density and independent of the wave-length), that this single event must be an ionization, or similar process, which has to take place inside a certain volume of only about 10 atomic-distances-cubed, in order to produce a specified mutation. According to our picture, the energy for overcoming the threshold must obviously be furnished by that explosion-like process, ionization or excitation. I call it explosion-like, because the energy spent in one ionization (spent, incidentally, not by the X-ray itself, but by a secondary electron it produces) is well known and has the comparatively enormous amount of 30 electron-volts. It is bound to be turned into enormously increased heat motion around the point where it is discharged and to spread from there in the form of a 'heat wave', a wave of intense oscillations of the atoms. That this heat wave should still be able to furnish the required threshold energy of 1 or 2 electron-volts at an average 'range of action' of about ten atomic distances, is not inconceivable, though it may well be that an unprejudiced physicist might have anticipated a slightly lower range of action. That in many cases the effect of the explosion will not be an orderly isomeric transition

X射线如何诱发突变

现在来看看X射线诱发的突变率。我们已经从育种实验中推断出：首先（根据突变率和剂量的比例），某个单一事件会产生突变；其次（根据定量的结果，以及突变率取决于累积的电离密度而同波长无关的事实），这种单一事件必定是一种电离作用或类似的过程，它还必须发生在边长大约只有10个原子距离的立方体之内，才能产生一个特定的突变。

根据这个描述，克服阈值的能量显然是一个类似爆炸的过程，诸如电离或激发过程所能提供的。我之所以将它称之为类似爆炸的过程，是因为一次电离作用消耗的能量（顺便说一下，并不是X射线本身消耗的，而是它产生的次级电子消耗的）是众所周知的，并且具有30电子伏的相对巨大的能量。它必然会在放电点周围转变为巨大的热运动，并从那里以“热浪”的形式扩散开来，即原子的强烈振荡波。在大约10个原子距离的平均作用范围内，这种热浪仍然能够提供所需的1或2电子伏的起始能量，这并不是不可想象的，尽管一个没有偏见的物理学家很可能会预期一个稍微低一些的作用范围。

66 but a lesion（损害）of the chromosome, a lesion that becomes lethal when, by ingenious crossings, the uninjured partner (the corresponding chromosome of the second set) is removed and replaced by a partner whose corresponding gene is known to be itself morbid—all that is absolutely to be expected and it is exactly what is observed.

THEIR EFFICIENCY DOES NOT DEPEND ON SPONTANEOUS MUTABILITY

Quite a few other features are, if not predictable from the picture, easily understood from it. For example, an unstable mutant does not on the average show a much higher X-ray mutation rate than a stable one. Now, with an explosion furnishing an energy of 30 electron-volts you would certainly not expect that it makes a lot of difference whether the required threshold energy is a little larger or a little smaller, say 1 or 1.3 volts.

在许多情况下，爆炸的效应将不是有序的异构跃迁，而是这一染色体的损伤。接着通过巧妙的杂交设计，当未受损伤的染色体（即第二套染色体，与受损伤的染色体配对的那一条）被移除，并被一个已知其相应基因本身病态的染色体取代时，这种损伤就会致命——所有这一切现象，绝对是可以预期的，也正是观察到的情况。

X射线的效率并不取决于自发突变性

还有很多别的特性，即使不能根据以上描述来预测，也是很容易理解的。例如，一个不稳定的突变体的 X 射线突变率，平均来说并不比那些稳定的突变体高很多。一次爆炸就可以提供30电子伏的能量，你肯定不会期望能量阈值大一点或小一点会有那么大的不同，比如说1或1.3伏。

REVERSIBLE MUTATIONS

In some cases a transition was studied in both directions, say from a certain 'wild' gene to a specified mutant and back from that mutant to the wild gene. In such cases the natural mutation rate is sometimes nearly the same, sometimes very different. At first sight one is puzzled, because the threshold to be overcome seems to be the same in both cases. But, of course, it need not be, because it has to be measured from the energy level of the starting configuration, and that may be different for the wild and the mutated gene. (See Fig. 12 on p. 54, where '1' might refer to the wild allele, '2' to the mutant, whose lower stability would be indicated by the shorter arrow.)

On the whole, I think, Delbrück's 'model' stands the tests fairly well and we are justified in using it in further considerations.

可逆突变

在有些情况下，可以沿两个方向来研究这一转换，比如从某个特定的“野生”基因到一个特定的突变体，再从这个突变体变回到野生基因。在这些情况下，自然突变率有时近乎相等，有时则很不相同。乍一看，人们会感到困惑，因为这两种情况下所要克服的阈值似乎是相等的。但是，当然，不必困惑，因为必须从初始构型的能级开始算起，这对于野生基因和突变基因可能不同。（见图12，可以用“1”表示野生型等位基因，“2”表示突变型等位基因，后者较低的稳定性将由较短的箭头表示。）

总之，我认为德尔布吕克的“模型”是经得起检验的，我们有充分的理由进一步考虑它的应用。

CHAPTER 6
Order, Disorder and Entropy

Nec corpus mentem ad cogitandum, nec mens corpus ad motum, neque ad quietem, nec ad aliquid (si quid est) aliud determinare potest.[1]

SPINOZA, *Ethics*, Pt III, Prop.2

A REMARKABLE GENERAL CONCLUSION FROM THE MODEL

67 Let me refer to the phrase on p. 62, in which I tried to explain that the molecular picture of the gene made it at least conceivable that the miniature code should be in one-to-one correspondence with a highly complicated and specified plan of development and should somehow contain the means of putting it into operation. Very well then, but how does it do this? How are we going to turn 'conceivability' into true understanding?

Delbrück's molecular model, in its complete generality, seems to contain no hint as to how the hereditary substance works. Indeed,

1. Neither can the body determine the mind to think, nor the mind determine the body to motion or rest or anything else (if such there be).

第六章
有序，无序和熵

躯体不能决定头脑的思考方式，头脑也休想决定躯体的运动状态。

——斯宾诺莎《伦理学》第三卷，命题2

模型提供的重要结论

让我引用第五章第7节（原文第62页）的最后一段话。在这段话中，我试图解释，根据基因的分子图，可以设想这一个微型密码应该同一个特定的高度复杂的发育蓝图有着一一对应的精确关系，并以某种方式蕴藏着使密码行使功能的手段。这很好，但它是如何做到的呢？我们如何把“想象力”转化为真正的认识呢？

德尔布吕克的分子模型具有完整的普遍性，似乎并不包含任何关于遗传物质是如何作用的提示。事实上，我并不指望物

I do not expect that any detailed information on this question is likely to come from physics in the near future. The advance is proceeding and will, I am sure, continue to do so, from biochemistry under the guidance of physiology and genetics.

No detailed information about the functioning of the genetical mechanism can emerge from a description of its structure so general as has been given above. That is obvious. But, strangely enough, 68 there is just one general conclusion to be obtained from it, and that, I confess, was my only motive for writing this book.

From Delbrück's general picture of the hereditary substance it emerges that living matter, while not eluding the 'laws of physics' as established up to date, is likely to involve 'other laws of physics' hitherto unknown, which, however, once they have been revealed, will form just as integral a part of this science as the former.

理学能够在近期内为这个问题提供详细的信息。生物化学正在发展，我确信，在生理学和遗传学指导下，还将继续发展。

关于遗传机制如何行使功能的详细信息，显然不能从上面所给出的如此笼统的结构描述中得到。但奇怪的是，由此却可以得出一个普遍的结论，我承认这正是我写这本书的唯一动机。

从德尔布吕克对遗传物质的一般描述可以看出，生命物质虽然不能逃避迄今已确立的“物理定律”，但很可能还涉及尚不为人所知的“其他物理定律”。然而，一旦这些未知定律被揭示出来，将和以前的物理定律一样，成为这门科学必不可少的组成部分。

ORDER BASED ON ORDER

This is a rather subtle line of thought, open to misconception in more than one respect. All the remaining pages are concerned with making it clear. A preliminary insight, rough but not altogether erroneous, may be found in the following considerations:

It has been explained in chapter 1 that the laws of physics, as we know them, are statistical laws.[1] They have a lot to do with the natural tendency of things to go over into disorder.

But, to reconcile the high durability of the hereditary substance with its minute size, we had to evade the tendency to disorder by 'inventing the molecule', in fact, an unusually large molecule which has to be a masterpiece of highly differentiated order, safeguarded by the conjuring（魔法）rod of quantum theory. The laws of chance are not invalidated by this 'invention', but their outcome is modified. The physicist is familiar with the fact that the classical laws of physics are modified by quantum theory, especially at low temperature. There are many instances of this. Life seems to be one of them, a particularly striking one. Life seems to be orderly and lawful behaviour of matter, not based exclusively on its tendency to go over from order to disorder, but based partly on existing order that is kept up.

To the physicist—but only to him—I could hope to make my

---

1. To state this in complete generality about 'the laws of physics' is perhaps challengeable. The point will be discussed in chapter 7.

有序来自有序

这是一个相当微妙的思路，而且容易导致多方面的误解。本书接下来的所有篇幅就是要理清这一思路。一个初步的认识，粗略但并非完全错误，可从如下几个方面得到：

第一章已经解释过，这些物理学定律，正如我们所知道的，都是统计学定律[1]。它们和事物向着无序发展的自然倾向密切相关。

但是，要协调遗传物质的高度持久性和它的微小体积，我们必须“发明那种分子”避开那无序的倾向。事实上，这是一种不寻常的特大分子，必须是高度分化的有序性的产物，受量子理论的魔棍保护。随机定律并没有因这种“发明”而失效，只是结果被修正了。物理学家都很熟悉这一事实，物理学经典定律已经被量子理论修正了，尤其在低温情况下。这样的例子很多。生命似乎就是其中之一，尤其引人注目。生命似乎是物质有序且有规律的行为，并不是完全倾向于从有序走向无序，而是部分基于维持现有的秩序。

对于这位物理学家来说——但也只是对德尔布吕克——

1. 要全面概括“物理学定律”，也许很具挑战性。这一点将在第七章中讨论。

69

view clearer by saying: The living organism seems to be a macroscopic system which in part of its behaviour approaches to that purely mechanical (as contrasted with thermodynamical) conduct to which all systems tend, as the temperature approaches absolute zero and the molecular disorder is removed.

The non-physicist finds it hard to believe that really the ordinary laws of physics, which he regards as the prototype of inviolable precision, should be based on the statistical tendency of matter to go over into disorder. I have given examples in chapter 1. The general principle involved is the famous Second Law of Thermodynamics (entropy principle) and its equally famous statistical foundation. On pp. 69–74 I will try to sketch the bearing of the entropy principle on the large-scale behaviour of a living organism—forgetting at the moment all that is known about chromosomes, inheritance, and so on.

我希望把我的观点说得更清楚一点：生命有机体似乎是一个宏观系统，随着温度趋近绝对零度，分子的无序性被消除，它的部分行为接近于所有系统趋向的纯力学行为（与之相对的是热力学行为）。

非物理学家会觉得难以置信，物理学的一般定律，即被他们认为是不可侵犯的精确性典范，应该基于物质趋向于无序的统计学倾向。我已经在第一章中给出了一些例子，涉及的普遍原理是著名的热力学第二定律（熵原理）及其同样著名的统计学基础。在本章第3-7节（原文第69-74页），我将试图概述熵原理在生命有机体的宏观行为中所起的作用——此刻请把一切染色体、遗传等有关的知识搁置脑后。

LIVING MATTER EVADES THE DECAY TO EQUILIBRIUM

What is the characteristic feature of life? When is a piece of matter said to be alive? When it goes on 'doing something', moving, exchanging material with its environment, and so forth, and that for a much longer period than we would expect of an inanimate piece of matter to 'keep going' under similar circumstances. When a system that is not alive is isolated or placed in a uniform environment, all motion usually comes to a standstill very soon as a result of various kinds of friction; differences of electric or chemical potential are equalized, substances which tend to form a chemical compound do so, temperature becomes uniform by heat conduction. After that the whole system fades away into a dead, inert（惰性的）lump of matter. A permanent state is reached, in which no observable events occur. The physicist calls this the state of thermodynamical equilibrium, or of 'maximum entropy'.

Practically, a state of this kind is usually reached very rapidly. Theoretically, it is very often not yet an absolute equilibrium, not yet the true maximum of entropy. But then the final approach to equilibrium is very slow. It could take anything between hours, years, centuries, …To give an example—one in which the approach is still fairly rapid: if a glass filled with pure water and a second one filled with sugared water are placed together in a hermetically（密封地）closed case at constant temperature, it appears at first that nothing happens, and the impression of complete equilibrium is created. But

生命物质与衰变的平衡

生命的典型特征是什么？什么时候一个物体可以说是活的？答案是当它在不停地“做什么”时，如移动，与环境进行物质交换，等等。这比那些处于相似环境中的非生命物质“保持运动”的时间要长得多。如果一个非生命的系统，将其隔离出来或者放在一个均一的环境之中，所有运动通常很快就会停止，这是各种摩擦力作用的结果；电势或化学势的差异被平衡，倾向于形成化合物的物质也是如此，温度也由于热传导而变得均一了。此后，整个系统便会衰退成一团死寂的、惰性的物质，达到一种持久不变的状态，观察不到任何可见变化。物理学家称这种状态为热力学平衡，或“最大熵”。

实际上，这种状态通常很快就能达到。从理论上说，这种状态往往不是绝对的平衡状态，也不是真正的最大熵。但是，最后趋于平衡的过程是非常缓慢的，也许是几小时、几年、几个世纪……

举一个例子，在这个例子中，这种趋向仍然是相当迅速的。如果一个玻璃杯盛满了清水，另一个玻璃杯盛满糖水溶液，把它们放入一个密封的恒温箱中，起初好像什么也没有发生，给

after a day or so it is noticed that the pure water, owing to its higher vapour pressure, slowly evaporates and condenses on the solution. The latter overflows. Only after the pure water has totally evaporated has the sugar reached its aim of being equally distributed among all the liquid water available.

These ultimate slow approaches to equilibrium could never be mistaken for life, and we may disregard them here. I have referred to them in order to clear myself of a charge of inaccuracy.

我们制造了完全平衡的假象。但过了一天左右的时间，可以看到，由于那杯清水的蒸气压较高，它慢慢地蒸发出来并凝结在另一杯糖水溶液液面上。糖水溶液满溢出来。只有当那杯清水全部蒸发完后，糖才均匀分布于恒温箱内所有的液态水之中。

这类最终缓慢地趋于平衡的过程永远不会被误认为是生命，对它们的讨论先到这儿。之所以讲到它们，是免得让人说我的话并不那么准确。

IT FEEDS ON ' NEGATIVE ENTROPY '

It is by avoiding the rapid decay into the inert state of 'equilibrium' that an organism appears so enigmatic; so much so, that from the earliest times of human thought some special non-physical or supernatural force (*vis viva*, entelechy) was claimed to be operative in the organism, and in some quarters is still claimed.

How does the living organism avoid decay? The obvious answer is: By eating, drinking, breathing and (in the case of plants) assimilating. The technical term is *metabolism*. The Greek word (μεταβαλλειν) means change or exchange. Exchange of what? Originally the underlying idea is, no doubt, exchange of material. (E.g. the German for metabolism is *Stoffwechsel*.) That the exchange of material should be the essential thing is absurd. Any atom of nitrogen, oxygen, sulphur, etc., is as good as any other of its kind; what could be gained by exchanging them? For a while in the past our curiosity was silenced by being told that we feed upon energy. In some very advanced country (I don't remember whether it was Germany 71 or the U.S.A. or both) you could find menu cards in restaurants indicating, in addition to the price, the energy content of every dish. Needless to say, taken literally, this is just as absurd. For an adult organism the energy content is as stationary as the material content. Since, surely, any calorie is worth as much as any other calorie, one cannot see how a mere exchange could help.

吃进“负熵”

正是由于能够避免迅速退化到“平衡”的惰性状态，生命体才显得如此神秘。这么说吧，从人类思想的最早期开始，曾认为某种特殊的、非物理的或超自然的力（*vis viva*，活力）在生命体中起着作用，现在仍然有人这样认为。

生命体是如何避免衰退的呢？显而易见的回答是：通过吃、喝、呼吸、消化（对植物来说称同化），专业术语叫“代谢”（*metabolism*）。这个希腊词（μεταβαλλειν）意为“变化”或“交换”。交换什么呢？最初的基本概念毫无疑问，是物质的交换（例如，德语中的代谢为*Stoffwechsel*）。物质交换是本质，这种看法是荒谬的。有机体中氮、氧、硫等任何一个原子和其他同类的原子都一样，相互交换又能得到什么呢？

在过去的一段时间里，我们不再好奇了，因为我们被告知我们依靠摄入能量而生存。在某些非常发达的国家（我不记得是德国还是美国，或者两国皆有），你可以在餐馆里找到菜单卡，除了价格之外，还标明每道菜含有的能量。不用说，从字面上看，这很荒谬。因为对于一个成年生命体来说，能量含量和物质含量都是固定不变的。毫无疑问，既然任何一个卡路里肯定跟其他卡路里的价值是一样的，那么，确实看不出单纯的交换究竟有什么好处。

What then is that precious something contained in our food which keeps us from death? That is easily answered. Every process, event, happening—call it what you will; in a word, everything that is going on in Nature means an increase of the entropy of the part of the world where it is going on. Thus a living organism continually increases its entropy—or, as you may say, produces positive entropy—and thus tends to approach the dangerous state of maximum entropy, which is of death. It can only keep aloof from it, i.e. alive, by continually drawing from its environment negative entropy—which is something very positive as we shall immediately see. What an organism feeds upon is negative entropy. Or, to put it less paradoxically, the essential thing in metabolism is that the organism succeeds in freeing itself from all the entropy it cannot help producing while alive.

那么，我们的食物中到底含有什么宝贵的东西，使我们免于死亡？这个问题很好回答。每个过程、事件、发生的事——无论你叫它什么；简言之，自然界中正在发生的一切都意味着在那个发生的区域的熵在增加。因此，一个活的生命体的熵在不断增加——或者，正如你说的，产生正熵——从而趋向于最大熵的危险状态，也就是死亡。要想远离最大熵活下去，只有从环境中不断汲取负熵——我们很快就会明白，负熵肯定是非常有用的东西。有机体吃进去的是负熵。或者，换一种不那么自相矛盾的说法：新陈代谢的本质就是，使有机体成功地让自己免受其生命活动产生的熵的影响。

WHAT IS ENTROPY?

What is entropy? Let me first emphasize that it is not a hazy（模糊的）concept or idea, but a measurable physical quantity just like of the length of a rod, the temperature at any point of a body, the heat of fusion of a given crystal or the specific heat of any given substance. At the absolute zero point of temperature (roughly —273°C) the entropy of any substance is zero. When you bring the substance into any other state by slow, reversible little steps (even if thereby the substance changes its physical or chemical nature or splits up into two or more parts be of different physical or chemical nature) the entropy increases by an amount which is computed by dividing every little portion of heat you had to supply in that procedure by the absolute temperature at which it was supplied—and by 72 summing up all these small contributions. To give an example, when you melt a solid, its entropy increases by the amount of the heat of fusion divided by the temperature at the melting-point. You see from this, that the unit in which entropy is measured is Cal./°C (just as the calorie is the unit of heat or the centimetre the unit of length).

熵是什么？

何为熵？我首先要强调，这并不是一种模糊的概念或观念，而是一个可度量的物理量，就像一根木棍的长度，或一个物体上任一点的温度，某个晶体的熔化热，或者任何一个物体的比热那样。温度处于绝对零度时（约为－273℃），任何物质的熵都是零。如果你把一个物体以缓慢的、可逆的微小步骤带到另一个状态（即使这一物体因而改变其物理性质或化学性质，或者分解成两块或好几块物理或化学性质都不相同的部分），这时熵的增量可以这样计算：将在此过程的每一小步必须提供的热量除以提供热量时的绝对温度，再把所有这些小的贡献累加起来。举一个例子，当你熔化一个固体时，其熵增就是熔化热除以熔点温度。由此可以看出，熵的单位是Cal./℃（就像卡路里是热量的单位，或厘米是长度的单位一样）。

THE STATISTICAL MEANING OF ENTROPY

I have mentioned this technical definition simply in order to remove entropy from the atmosphere of hazy mystery that frequently veils it. Much more important for us here is the bearing on the statistical concept of order and disorder, a connection that was revealed by the investigations of Boltzmann and Gibbs in statistical physics. This too is an exact quantitative connection, and is expressed by

$$\text{entropy} = k\log D,$$

where k is the so-called Boltzmann constant ($=3.2983 \times 10^{-24}$ Cal./°C), and D a quantitative measure of the atomistic disorder of the body in question. To give an exact explanation of this quantity D in brief non-technical terms is well-nigh impossible. The disorder it indicates is partly that of heat motion, partly that which consists in different kinds of atoms or molecules being mixed at random, instead of being neatly separated, e.g. the sugar and water molecules in the example quoted above. Boltzmann's equation is well illustrated by that example. The gradual 'spreading out' of the sugar over all the water available increases the disorder D, and hence (since the logarithm of D increases with D) the entropy. It is also pretty clear that any supply of heat increases the turmoil of heat motion, that is to say, increases D and thus increases the entropy; it is particularly clear that this should be so when you melt a crystal, since you thereby destroy the neat and permanent arrangement of the atoms or molecules and turn the crystal lattice into a continually changing random

熵的统计学意义

说到熵这个专业术语的定义，我只不过是为了揭开时常笼罩在它周围的神秘面纱。对我们来说，更为重要的是熵与有序和无序这一统计学概念的联系，这种联系是玻尔兹曼和吉布斯在统计物理学方面的研究揭示的，这也是一种精确的定量联系，可以表达为：

$$熵 = k\log D,$$

这里k是所谓的玻尔兹曼常数（$=3.2983 \times 10^{-24}$ Cal./°C），D是对所讨论物体原子无序性的定量度量。用简短的非专业术语对D这个量做出精确的解释几乎是不可能的。它所表示的无序，一部分是热运动的无序，另一部分是来自不同种类的原子或分子的随机混合，而不是整齐地分开，如同前面引用例子中的糖分子和水分子。这个例子可以很好地解读玻尔兹曼方程。糖逐渐“分布”到整杯水中，增加了系统的无序性D，因此（因为D的对数随着D的增加而增加）增加了熵。同样清楚的是，提供任何热量都会增加热运动的混乱程度，也就是说增加了D，进而也增加了熵。特别清楚的是，当你熔化一个晶体时，因为你破坏了原子或分子整齐而稳定的排列，晶体晶格变成了一种连续变化的随机分布。

distribution.

An isolated system or a system in a uniform environment (which 73 for the present consideration we do best to include as a part of the system we contemplate) increases its entropy and more or less rapidly approaches the inert state of maximum entropy. We now recognize this fundamental law of physics to be just the natural tendency of things to approach the chaotic state (the same tendency that the books of a library or the piles of papers and manuscripts on a writing desk display) unless we obviate（避免）it. (The analogue of irregular heat motion, in this case, is our handling those objects now and again without troubling to put them back in their proper places.)

一个孤立的系统或者处于均一环境中的系统（就目前的考虑而言，我们尽量把这个环境作为我们所考察的系统的一部分），它的熵总是增大的，而且或早或晚会趋向最大熵的惰性状态。我们现在认识到，这一物理学基本定律就是，事物会自然地趋向混乱状态（就像图书馆里的书或写字台上的一大叠纸张和手稿那样），除非我们进行干预（在这种情况下，与不规则热运动类似的是，我们不时地去搬动那些东西，但又不肯花点力气把它们放回原处）。

ORGANIZATION MAINTAINED BY EXTRACTING 'ORDER' FROM THE ENVIRONMENT

How would we express in terms of the statistical theory the marvellous faculty of a living organism, by which it delays the decay into thermodynamical equilibrium (death)? We said before: 'It feeds upon negative entropy', attracting, as it were, a stream of negative entropy upon itself, to compensate the entropy increase it produces by living and thus to maintain itself on a stationary and fairly low entropy level.

If D is a measure of disorder, its reciprocal, $1/D$, can be regarded as a direct measure of order. Since the logarithm of $1/D$ is just minus the logarithm of D, we can write Boltzmann's equation thus:

$$-(\text{entropy}) = k \log (1/D).$$

Hence the awkward expression 'negative entropy' can be replaced by a better one: entropy, taken with the negative sign, is itself a measure of order. Thus the device by which an organism maintains itself stationary at a fairly high level of orderliness (= fairly low level of entropy) really consists in continually sucking orderliness from its environment. This conclusion is less paradoxical than it appears at first sight. Rather could it be blamed for triviality. Indeed, in the case of higher animals we know the kind of orderliness they feed upon well enough, viz. the extremely well-ordered state of matter in more or less complicated organic compounds, which serve them as 74 foodstuffs. After utilizing it they return it in a very much degraded form—not entirely degraded, however, for plants can still make use of it. (These, of course, have their most power supply of 'negative entropy' in the sunlight.)

借外"序"以稳内序

我们如何用统计学理论的语言来表达一个生命有机体的神奇能力，通过它延缓衰变到热力学平衡（死亡）呢？如前所说，"生命体以负熵为食"，也就是说它本身就吸引了一股负熵流，以补偿它在生命活动中产生的熵增，使自己保持在稳定而又低熵的水平。

如果D是对无序性的度量，它的倒数$1/D$，就可以看成有序性的直接量值。因为$1/D$的对数正好就是D的负对数，我们可以将玻尔兹曼方程写成：

$$\text{负熵} = k\log(1/D)$$

因此，"负熵"这个别扭的表述，就可以有一个更好的表达：熵，带有负号，其本身就是对有序性的度量。因此，有机体用于使自身维持在一个相对高水平的有序状态（等于较低水平的熵）的机制，实际上在于连续不断地从环境中汲取"有序"。这一结论乍看上去不再是那么自相矛盾，不过可能因为太通俗而受到批评。事实上，在高等动物的例子中，我们知道它们完全以汲取这样的"有序"为食，即在极为有序状态下的或复杂或简单的有机化合物。高等动物在享用了这些食物以后，排泄出来的则是高度降解了的物质——然而还不是彻底的降解，因为植物仍能利用它们（当然，植物最强大的"负熵"蕴藏在阳光之中）。

NOTE TO CHAPTER 6

The remarks on *negative entropy* have met with doubt and opposition from physicist colleagues. Let me say first, that if I had been catering for them alone I should have let the discussion turn on *free energy* instead. It is the more familiar notion in this context. But this highly technical term seemed linguistically too near to *energy* for making the average reader alive to the contrast between the two things. He is likely to take *free* as more or less an *epitheton ornans* (修饰语) without much relevance, while actually the concept is a rather intricate one, whose relation to Boltzmann's order-disorder principle is less easy to trace than for entropy and 'entropy taken with a negative sign', which by the way is not my invention. It happens to be precisely the thing on which Boltzmann's original argument turned.

But F. Simon has very pertinently pointed out to me that my simple thermodynamical considerations cannot account for our having to feed on matter 'in the extremely well ordered state of more or less complicated organic compounds' rather than on charcoal or diamond pulp. He is right. But to the lay reader I must explain that a piece of un-burnt coal or diamond, together with the amount of oxygen needed for its combustion, is also in an extremely well ordered state, as the physicist understands it. Witness to this: if you allow the reaction, the burning of the coal, to take place, a great amount of heat is produced. By giving it off to the surroundings, the system disposes of the very considerable entropy increase entailed by the reaction, and reaches a state in which it has, in point of fact, roughly the same entropy as before.

本章的注释

关于负熵的说法遭到了物理学同行们的质疑和反对。首先我想说，如果只是想迎合他们，我早就转而讨论自由能了。在这个情景下，自由能是更为人所熟知的概念。但是，这个高度专业的术语在语言学上似乎和“能量”太相似了，现今普通的读者难以区分两者的差别。读者很可能会把“自由”当作一个或多或少不那么相关的修饰词，而实际上，“自由能”这个概念相当微妙，它与玻尔兹曼有序-无序原理的关系，比熵和“带负号的熵”的关系更不容易理解，顺便说一句，这个概念并不是我发明的。碰巧的是，它正好是玻尔兹曼最初论证的出发点。

但西蒙非常中肯地向我指出，我那些简单的热力学思考还不足以说明这样的现象，即为何我们赖以为生的物质是“在极为有序状态下的或复杂或简单的有机化合物”，而不是木炭或钻石。他是对的。但对于外行读者，我必须解释一下，一块未燃烧的煤或金刚石连同燃烧时所需的氧气，都处于极为有序的状态，正如物理学家所理解的那样。要证明这一点：如果允许这一反应——煤的燃烧——发生，就会产生大量的热。通过将热量释放到周围环境中，系统便消除了反应带来的大量熵增，并达到了一种状态，实际上，它的熵与之前大致相同。

Yet we could not feed on the carbon dioxide that results from the reaction. And so Simon is quite right in pointing out to me, as he did, that actually the energy content of our food *does* matter; so my mocking at the menu cards that indicate it was out of place. Energy is needed to replace not only the mechanical energy of our bodily exertions, but also the heat we continually give off to the environment. And that we give off heat is not accidental, but essential. For this is precisely the manner in which we dispose of the surplus entropy we continually produce in our physical life process.

This seems to suggest that the higher temperature of the warm-blooded animal includes the advantage of enabling it to get rid 75 of its entropy at a quicker rate, so that it can afford a more intense life process. I am not sure how much truth there is in this argument (for which I am responsible, not Simon). One may hold against it, that on the other hand many warm-blooders are *protected* against the rapid loss of heat by coats of fur or feathers. So the parallelism between body temperature and 'intensity of life', which I believe to exist, may have to be accounted for more directly by van't Hoff's law, mentioned on p. 65: the higher temperature itself speeds up the chemical reactions involved in living. (That it actually does, has been confirmed experimentally in species which take the temperature of the surroundings.)

然而，我们并不能以这一反应产生的二氧化碳为食。所以，西蒙非常正确地向我指出：事实上，我们的食物中所含的能量确实很重要，因而我对菜单上标出食物所含能量的嘲讽并不合适。我们不仅需要能量来提供身体活动所需的机械能量，而且需要它补充由身体不断释放到环境中的热量。我们散发热量并不是偶然的，而是必不可少的。因为我们正是以这种方式来清除在生命活动过程中不断产生的多余的熵。

这似乎表明，温血动物拥有较高的体温这一优势，能以较快的速度排出身体产生的熵，因此能够承受强度更大的生命活动。我并不能肯定这个观点在多大程度上符合实情（对此负责的是我，而不是西蒙）。有人可能会反对这种观点，因为另一方面，有许多温血动物以皮毛或羽毛来防止热的快速散失。

因此，体温同“生命强度”之间的对应关系，我认为是存在的，可能必须由范托夫定律来更为直接地予以解释，在第五章第11节（原文第65页）提到：正是较高的温度加速了生命活动中的化学反应（事实的确如此，这一点已经通过实验在那些从环境中吸收热量的物种中得到了证实）。

CHAPTER 7
Is Life Based on the Laws of Physics?

Si un hombre nunca se contradice, será porque nunca dice nada.[1]

MIGUEL DE UNAMUNO (quoted from conversation)

NEW LAWS TO BE EXPECTED IN THE ORGANISM

76 What I wish to make clear in this last chapter is, in short, that from all we have learnt about the structure of living matter, we must be prepared to find it working in a manner that cannot be reduced to the ordinary laws of physics. And that not on the ground that there is any 'new force' or what not, directing the behaviour of the single atoms within a living organism, but because the construction is different from anything we have yet tested in the physical laboratory. To put it crudely, an engineer, familiar with heat engines only, will, after inspecting the construction of an electric motor, be prepared to find it working along principles which he does not yet understand.

1. If a man never contradicts himself, the reason must be that he virtually never says anything at all.

第七章
生命遵循物理定律吗？

开口，必有矛盾之处，闭嘴，当然决无此忧。

——乌纳穆诺（引自对话）

生命体中可能存在新的定律

在最后一章里，我想阐明的是，简而言之，基于我们对生命物质结构的所有了解，我们必须做好准备，去发现生命活动以一种不可能被简化成普通物理学定律的方式运行。这并不是因为是否有任何一种“新奇的力”，指导着生命有机体中各个单原子的行为，而是因为它的结构与迄今为止我们在物理学实验室中测试过的所有实验材料都不同。

举个不大确切的例子：一名只熟悉热机的工程师，在检查完电动机的构造之后，一定会发现电动机是按照他尚不了解的工作原理运转的。他会发现他所熟悉的用于制作水壶的铜，在

He finds the copper familiar to him in kettles used here in the form of long, long wires wound in coils; the iron familiar to him in levers and bars and steam cylinders is here filling the interior of those coils of copper wire. He will be convinced that it is the same copper and the same iron, subject to the same laws of Nature, and he is right in that. The difference in construction is enough to prepare him for an entirely different way of functioning. He will not suspect that an electric motor is driven by a ghost because it is set spinning by the turn of a switch, without boiler and steam.

这里成了绕成一圈圈的、长长的铜线；他熟悉的用作杠杆、栏杆以及蒸汽缸的铁，在这里被插入铜线圈里作内芯。但他确信，铜还是同样的铜，铁还是同样的铁，都遵循相同的自然定律。在这一点上他是对的。这些构造的不同，足以使他接受一种完全不同的运作方式。他不会因为电动机没有锅炉和蒸汽，是由开关转动而旋转的，就怀疑电动机是由幽灵驱动的。

REVIEWING THE BIOLOGICAL SITUATION

77 The unfolding of events in the life cycle of an organism exhibits an admirable regularity and orderliness, unrivalled by anything we meet with in inanimate matter. We find it controlled by a supremely well-ordered group of atoms, which represent only a very small fraction of the sum total in every cell. Moreover, from the view we have formed of the mechanism of mutation we conclude that the dislocation of just a few atoms within the group of 'governing atoms' of the germ cell suffices to bring about a well-defined change in the large-scale hereditary characteristics of the organism.

These facts are easily the most interesting that science has revealed in our day. We may be inclined to find them, after all, not wholly unacceptable. An organism's astonishing gift of concentrating a 'stream of order' on itself and thus escaping that the decay into atomic chaos—of 'drinking orderliness' from a suitable environment—seems to be connected with the presence of the 'aperiodic solids', the chromosome molecules, which doubtless represent the highest degree of well-ordered atomic association we know of—much higher than the ordinary periodic crystal—in virtue of the individual role every atom and every radical is playing here.

To put it briefly, we witness the event that existing order displays the power of maintaining itself and of producing orderly events. That sounds plausible enough, though in finding it plausible we, no doubt, draw on experience concerning social organization and other events which involve the activity of organisms. And so it might seem that something like a vicious circle is implied.

生物学进展概述

在一个有机体的生命周期中展开的一个又一个事件，表现出一种令人敬佩的规律性和秩序性，这是我们在非生命物质中所遇到的任何事物所无法比拟的。我们发现，生命由一种极其有序的原子团控制，这些原子只占每个细胞中原子总数的很小一部分。此外，根据已经形成的关于突变机制的认识，我们得出这样的结论：在生殖细胞的“支配性原子”团里，只要那么几个原子错位，就足以使有机体的宏观遗传性状发生改变。

这些无疑是当代科学揭示的最有趣的事实。毕竟，我们或许更倾向于认为它们并非完全不能接受。生命体将“有序之流”集中于它自身，从而避免衰变为混沌状态的原子——从合适的环境中“吸取有序”——的惊人的天赋，似乎与“非周期性固体”，即染色体分子的存在有关。毫无疑问，这些分子代表着已知最高级别有序的原子集合——要比普通的周期性晶体级别高得多——由于每一个原子和每一个基团都在这里发挥各自的作用。

简言之，我们见证了现有秩序显示出维持自身和产生有序事件的能力。这种说法听起来很有道理，尽管为了说明它是合理的，毫无疑问，我们吸取了关于社会组织以及与有机体活动有关的其他事件的经验。因此，这似乎有点暗示了一种恶性循环。

SUMMARIZING THE PHYSICAL SITUATION

However that may be, the point to emphasize again and again is that to the physicist the state of affairs is not only not plausible but most exciting, because it is unprecedented. Contrary to the common 78 belief, the regular course of events, governed by the laws of physics, is never the consequence of one well-ordered configuration of atoms—not unless that configuration of atoms repeats itself a great number of times, either as in the periodic crystal or as in a liquid or in a gas composed of a great number of identical molecules.

Even when the chemist handles a very complicated molecule *in vitro* he is always faced with an enormous number of like molecules. To them his laws apply. He might tell you, for example, that one minute after he has started some particular reaction half of the molecules will have reacted, and after a second minute three-quarters of them will have done so. But whether any particular molecule, supposing you could follow its course, will be among those which have reacted or among those which are still untouched, he could not predict. That is a matter of pure chance.

This is not a purely theoretical conjecture. It is not that we can never observe the fate of a single small group of atoms or even of a single atom. We can, occasionally. But whenever we do, we find complete irregularity, co-operating to produce regularity only on the average. We have dealt with an example in chapter 1. The Brownian movement of a small particle suspended in a liquid is completely ir-

物理学进展概述

无论如何，我们需要再度反复强调的一点是，对于物理学家来说，这一进展既难以置信，又是那么令人兴奋，因为它是前所未有的。与普遍观念相反，受物理定律支配的事件的有序进程，从来就不是原子的高度有序的构型所导致的结果——除非原子构型不断重复，要么在周期性晶体中，要么在液体或气体中，它们由不计其数的同样的分子组成。

即使当那个化学家在试管中研究一种非常复杂的分子时，他面对的也是无数相似的分子，可见化学定律也适用于这些分子。他也许会告诉你，例如，在某个特定的反应开始一分钟后，一半的分子会发生反应，第二分钟之后，会有四分之三的分子发生反应。但是无论任何特定的分子，假设你可以追踪，它的运动轨迹是在已经起反应的分子中，还是在未起反应的分子中，他无法预测，这是纯粹偶然的。

这并不是一种纯理论的推测，也并不是说我们永远无法观察到一小簇原子团，甚至是单原子的命运。我们偶尔能够做到。但无论何时我们这样做，我们都会发现完全的不规则性，它们只在平均水平上相互作用产生规则性。我们曾在第一章讨论过

regular. But if there are many similar particles, they will by their irregular movement give rise to the regular phenomenon of diffusion.

The disintegration of a single radioactive atom is observable (it emits a projectile which causes a visible scintillation on a fluorescent screen). But if you are given a single radioactive atom, its probable lifetime is much less certain than that of a healthy sparrow. Indeed, nothing more can be said about it than this: as long as it lives (and that may be for thousands of years) the chance of its blowing up within the next second, whether large or small, remains the same. This patent lack of individual determination nevertheless results in the exact exponential law of decay of a large number of radioactive atoms of the same kind.

一个例子。悬浮在液体中的一颗微粒的布朗运动是完全不规则的。但如果有许多相似的微粒，它们的不规则运动将会引起规则的扩散现象。

单个放射性原子的衰变是可观察的（它会释放出抛射体，在荧光屏上产生可见的闪烁）。但如果给你一个放射性原子，它的寿命可能要比一只健康的麻雀更不确定。事实上，关于它没有什么可说的了：只要它活着（甚至可能活了几千年），它在下一秒钟里爆炸的可能性，无论是大是小，总保持不变。然而，这种缺乏个体决定的能力，导致了大量同类放射性原子衰变的精确指数定律。

THE STRIKING CONTRAST

79 In biology we are faced with an entirely different situation. A single group of atoms existing only in one copy produces orderly events, marvellously tuned in with each other and with the environment according to most subtle laws. I said, existing only in one copy, for after all we have the example of the egg and of the unicellular organism. In the following stages of a higher organism the copies are multiplied, that is true. But to what extent? Something like 10^{14} in a grown mammal, I understand. What is that! Only a millionth of the number of molecules in one cubic inch of air. Though comparatively bulky, by coalescing they would form but a tiny drop of liquid. And look at the way they are actually distributed. Every cell harbours just one of them (or two, if we bear in mind diploidy). Since we know the power this tiny central office has in the isolated cell, do they not resemble stations of local government dispersed through the body, communicating with each other with great ease, thanks to the code that is common to all of them?

Well, this is a fantastic description, perhaps less becoming a scientist than a poet. However, it needs no poetical imagination but only clear and sober scientific reflection to recognize that we are here obviously faced with events whose regular and lawful unfolding is guided by a 'mechanism' entirely different from the 'probability mechanism' of physics. For it is simply a fact of observation that the guiding principle in every cell is embodied in a single atomic asso-

显著的对比

在生物学中，我们面临着完全不同的情况。仅有一个拷贝的一簇原子团会产生有序的事件，它们彼此之间以及与环境之间按照最微妙的规律奇妙地协调着。

我之所以说仅有一份拷贝，是因为我们毕竟有卵细胞和单细胞有机体这样的例子。在高等生物的后续阶段，拷贝会成倍增加，这倒是真的。但是会增加到什么程度呢？按我的理解，在成年哺乳动物中大约是10^{14}个。那到底有多少呢？相当于1立方英寸空气中分子总数的百万分之一。虽然相对来说体积较大，但是聚集起来只不过形成一个很小的液滴。再看看它们的实际分布，每一个细胞中刚好只有一份拷贝（或者两份，如果我们心里想的是二倍体）。既然我们已经知道这个小小的中央机关在单个细胞中所拥有的权利，它们不就像遍布全身的“地方政府机构”吗？它们之间的相互沟通非常方便，这多亏所有这些机构都有共同的密码。

是啊，真是一个绝妙的描述，也许更像一位诗人而非科学家了。然而，无须诗意的想象，只需清晰而清醒的科学思维，就可以认识到我们在这里面对的显然是这样一些事件，它们是由一种完全不同于物理学“概率机制”的“机制”指导，有规律有序地展开。因为这仅仅是观察到的事实，即每一个细胞的指导原则都蕴藏在一个原子集合体之中，它只有一个拷贝（有时有

ciation existing only one copy (or sometimes two)—and a fact of observation that it results in producing events which are a paragon of orderliness. Whether we find it astonishing or whether we find it quite plausible that a small but highly organized group of atoms be capable of acting in this manner, the situation is unprecedented, it is unknown anywhere else except in living matter. The physicist and the chemist, investigating inanimate matter, have never witnessed phenomena which they had to interpret in this way. The case did not arise and so our theory does not cover it—our beautiful statistical theory of which we were so justly proud because it allowed us to look behind the curtain, to watch the magnificent order of exact physical law coming forth from atomic and molecular disorder; because it revealed that the most important, the most general, the all-embracing law of entropy could be understood without a special assumption *ad hoc*（特别的）, for it is nothing but molecular disorder itself.

两个拷贝）—— 而且直接观察到的事实是，它可能导致井然有序的典范事件。不管我们是否感到惊讶，或者我们是否觉得相当合理，一个小而高度组织的原子团能够以这种方式活动，这种情形是前所未见的，也是任何别的事物都未曾有过的，除了有机体以外。物理学家和化学家在研究非生命物质时，从未见过他们必须按照这种方式来解释的现象。

这种事件前所未有，所以我们的理论未曾涉及 —— 我们完美的统计学理论之所以如此值得我们自豪，是因为它使我们看到了幕后，看到了从原子和分子的无序中产生了精确物理定律的奇妙有序；是因为它揭示了最重要，最普遍，无处不在的熵定律，不需要特别的假设就可以理解，因为熵并非别的什么东西，只不过是分子自身的无序罢了。

TWO WAYS OF PRODUCING ORDERLINESS

The orderliness encountered in the unfolding of life springs from a different source. It appears that there are two different 'mechanisms' by which orderly events can be produced: the 'statistical mechanism' which produces 'order from disorder' and the new one, producing 'order from order'. To the unprejudiced mind the second principle appears to be much simpler, much more plausible. No doubt it is. That is why physicists were so proud to have fallen in with the other one, the 'order-from-disorder' principle, which is actually followed in Nature and which alone conveys an understanding of the great line of natural events, in the first place of their irreversibility. But we cannot expect that the 'laws of physics' derived from it suffice straightaway to explain the behaviour of living matter, whose most striking features are visibly based to a large extent on the 'order-from-order' principle. You would not expect two entirely different mechanisms to bring about the same type of law—you would not expect your latch-key, to open your neighbour's door as well.

We must therefore not be discouraged by the difficulty of interpreting life by the ordinary laws of physics. For that is just what is to be expected from the knowledge we have gained of the structure of living matter. We must be prepared to find a new type of physical law prevailing in it. Or are we to term it a non-physical, not to say a super-physical, law?

有序的两种产生方式

在生命过程中表现出来的有序来源不同。似乎有两种不同的“机制”可以产生有序事件：“有序来自无序”的“统计学机制”和“有序来自有序”这一新的机制。在没有偏见的人看来，第二条原理似乎更为简单，更加可信，这是毫无疑问的。这就是为什么物理学家们如此引以为傲地发现了另一种原理，即“有序来自无序”原理。自然界实际遵循着这条原理，而且只有它才使我们理解了自然事件的伟大脉络，首先是自然事件的不可逆性。但是我们不能指望由此推导的“物理定律”足以解释生命物质的行为，因为生命物质行为最显著的特征，显然在很大程度上基于“有序来自有序”这一原理。你不能指望两种完全不同的机制会归纳出同一种定律——就像你不能指望用你的钥匙也能打开邻居家的门锁。

因此，我们不必因为用普通的物理定律难以解释生命现象而感到沮丧。这正是根据我们对生命物质结构的已有认识而预计得到的。我们必须做好准备，从中寻找一种新的物理定律。或者，我们称它为“非物理”，乃至“超物理”定律呢？

THE NEW PRINCIPLE IS NOT ALIEN TO PHYSICS

81 No. I do not think that. For the new principle that is involved is a genuinely physical one: it is, in my opinion, nothing else than the principle of quantum theory over again. To explain this, we have to go to some length, including a refinement, not to say an amendment, of the assertion previously made, namely, that all physical laws are based on statistics.

This assertion, made again and again, could not fail to arouse contradiction. For, indeed, there are phenomena whose conspicuous features are visibly based directly on the 'order-from-order' principle and appear to have nothing to do with statistics or molecular disorder.

The order of the solar system, the motion of the planets, is maintained for an almost indefinite time. The constellation of this moment is directly connected with the constellation at any particular moment in the times of the Pyramids; it can be traced back to it, or vice versa. Historical eclipses have been calculated and have been found in close agreement with historical records or have even in some cases served to correct the accepted chronology. These calculations do not imply any statistics, they are based solely on Newton's law of universal attraction.

新定律与物理学并不相悖

不，我不这么认为。因为其中涉及的新原理是不折不扣的物理学定律：在我看来，这只不过是量子理论原理的再次演绎罢了。要想解释这一点，我们不得不多说几句，可以说是对先前的论断的细化，抑或是补充，关于一切物理定律都是建立在统计基础上的。

这个一再被提起的论断，不可避免地会带来矛盾。因为，确实，有一些现象的显著特征显然是直接基于“有序来自有序”原理，似乎与统计学和分子无序性毫不相关。

太阳系的秩序，行星的运转，几乎亘古不变。此时此刻的星座，与金字塔时代任一时刻的星座直接相关，使我们可以据此追溯到那个时候，反之亦然。历史上的日食月食已经被计算出来，并且发现与实际记录居然十分吻合，甚至在某些情况下，还被用来校正公认的年表。这些计算不涉及任何统计学，完全基于牛顿的万有引力定律。

Nor does the regular motion of a good clock or any similar mechanism appear to have anything to do with statistics. In short, all purely mechanical events seem to follow distinctly and directly the 'order-from-order' principle. And if we say 'mechanical', the term must be taken in a wide sense. A very useful kind of clock is, as you know, based on the regular transmission of electric pulses from the power station.

I remember an interesting little paper by Max Planck on the topic 'The Dynamical and the Statistical Type of Law' ('Dynamische und Statistische Gesetzmässigkeit'). The distinction is precisely the one we have here labelled as 'order from order' and 'order from disorder'. The object of that paper was to show how the interesting statistical type of law, controlling large-scale events, is constitut-82 ed from the 'dynamical' laws supposed to govern the small-scale events, the interaction of the single atoms and molecules. The latter type is illustrated by large-scale mechanical phenomena, as the motion of the planets or of a clock, etc.

Thus it would appear that the 'new' principle, the order-from-order principle, to which we have pointed with great solemnity as being the real clue to the understanding of life, is not at all new to physics. Planck's attitude even vindicates priority for it. We seem to arrive at the ridiculous conclusion that the clue to the understanding of life is that it is based on a pure mechanism, a 'clock-work' in the sense of Planck's paper, The conclusion is not ridiculous and is, in my opinion, not entirely wrong, but it has to be taken 'with a very big grain of salt'.

一个精准的时钟，或者任何类似的机械装置的规则运动，似乎与统计学也同样毫不相关。简而言之，所有纯粹的机械事件似乎都明确而直接地遵循“有序来自有序”原理。如果我们说什么是“机械的”，这个词必须从广义上来理解。一个非常有用的时钟，如你所知，是靠发电站定期地输送电脉冲来运转的。

我记得马克斯·普朗克写过一篇很有意思的小文章，题目是“动力学和统计学定律”。这两者的区别正好就是我们所谓的“有序来自有序”和“有序来自无序”的区别。那篇文章旨在表明控制宏观事件的统计学定律如何有趣，是如何由控制微观事件的动力学定律构成的，即单原子和单分子相互作用。后一种类型可以用宏观的机械现象来说明，如行星或时钟的运动等。

由此可见，我们庄严地指出的“新原理”，即“有序来自有序”，是理解生命的真正线索，对物理学来说根本不是新鲜事物。普朗克的态度甚至证明了此原理的优越性。我们似乎得出了一个荒谬的结论：理解生命的线索是基于一种纯粹的机械论，即普朗克文章中说的那种“时钟式运转”。在我看来，这个结论倒是没那么荒谬，也不是完全错误的，但必须持“高度保留”的态度去看待。

THE MOTION OF A CLOCK

Let us analyse the motion of a real clock accurately. It is not at all a purely mechanical phenomenon. A purely mechanical clock would need no spring, no winding. Once set in motion, it would go on forever. A real clock without a spring stops after a few beats of the pendulum, its mechanical energy is turned into heat. This is an infinitely complicated atomistic process. The general picture the physicist forms of it compels him to admit that the inverse process is not entirely impossible: a springless clock might suddenly begin to move, at the expense of the heat energy of its own cog wheels and of the environment. The physicist would have to say: The clock experiences an exceptionally intense fit of Brownian movement. We have seen in chapter 2 (p. 16) that with a very sensitive torsional balance (electrometer or galvanometer) that sort of thing happens all the time. In the case of a clock it is, of course, infinitely unlikely.

Whether the motion of a clock is to be assigned to the dynamical or to the statistical type of lawful events (to use Planck's expressions) depends on our attitude. In calling it a dynamical phenomenon we fix attention on the regular going that can be secured by a 83 comparatively weak spring, which overcomes the small disturbances by heat motion, so that we may disregard them. But if we remember that without a spring the clock is gradually slowed down by friction, we find that this process can only be understood as a statistical phenomenon.

时钟的运动

让我们准确地分析一下一个真正的时钟运动。它绝不是一种纯粹的机械现象。一个纯粹的机械钟应该不需要发条和弹簧。它一旦开始运动，就会永远走下去。一个没有发条的真的时钟的钟摆摆了那么几下就会停下来，它的机械能被转化为热能，这是一种无限复杂的原子过程。物理学家对它的一般性描述迫使他承认，逆向过程不是完全不可能的：一台没有发条的时钟可能会突然开始走动，依靠的是它自身齿轮和环境的热能。这位物理学家一定会说：这个时钟经历了一次异常强烈的布朗运动。我们在第一章（第9节，原文第16页）里已经看到了，用一种非常灵敏的扭力天平（静电计或检流计）就总能观察到这类现象。当然，对于时钟来说，这永远是不可能的。

时钟的运动到底应当被归为动力学类型还是统计学类型（按普朗克的说法），取决于我们的态度。把它称为动力学现象，我们的关注点在于，它的规则运动用一根比较松散的发条就能驱动，只要克服热运动带来的微小干扰就行了，所以可以忽略不计。但是，如果我们还记得，没有了发条，时钟就会由于摩擦力而逐渐变慢，我们就会发现这个过程只能被理解为一种统计学现象。

However insignificant the frictional and heating effects in a clock may be from the practical point of view, there can be no doubt that the second attitude, which does not neglect them, is the more fundamental one, even when we are faced with the regular motion of a clock that is driven by a spring. For it must not be believed that the driving mechanism really does away with the statistical nature of the process. The true physical picture includes the possibility that even a regularly going clock should all at once invert its motion and, working backward, rewind its own spring—at the expense of the heat of the environment. The event is just 'still a little less likely' than a 'Brownian fit' of a clock without driving mechanism.

然而，时钟中的摩擦效应和热效应是微不足道的，这也许是从实际角度出发，无疑还会有第二种观点，它没有忽视这些效应，这是更为基本的一种观点，甚至当我们面对一台以发条为动力的时钟规则运动时也是如此。因为，绝不能认为驱动机制真的与这一过程的统计学性质无关。真正的物理学描述包括这样一种可能性：即使是一台正常运行的时钟，也可能突然逆转它的运动，向后倒退，重新上紧自己的发条——以牺牲环境热能为代价。这种事件的可能性只是比没有驱动机制的时钟的“异常强烈的布朗运动”的可能性小一些。

CLOCKWORK AFTER ALL STATISTICAL

Let us now review the situation. The 'simple' case we have analysed is representative of many others—in fact of all such as appear to evade the all-embracing principle of molecular statistics. Clockworks made of real physical matter (in contrast to imagination) are not true 'clock-works'. The element of chance may be more or less reduced, the likelihood of the clock suddenly going altogether wrong may be infinitesimal, but it always remains in the background. Even in the motion of the celestial bodies irreversible frictional and thermal influences are not wanting. Thus the rotation of the earth is slowly diminished by tidal friction, and along with this reduction the moon gradually recedes from the earth, which would not happen if the earth were a completely rigid rotating sphere.

Nevertheless the fact remains that 'physical clock-works' visibly display very prominent 'order-from-order' features—the type that aroused the physicist's excitement when he encountered them in 84 the organism. It seems likely that the two cases have after all something in common. It remains to be seen what this is and what is the striking difference which makes the case of the organism after all novel and unprecedented.

钟表运动源于统计学

现在让我们归纳一下这个情况。我们所分析过的“简单”事件对很多其他事件来说都非常具有代表性，事实上，所有这些事件似乎都回避了分子统计学包罗万象的原理。由真实的物质（与我们想象的不同）制成的钟表并非真正的“时钟式运转”。尽管偶然性的因素也许多多少少能得到消减，时钟突然完全走错的可能性也微乎其微，但是它们总还是存在的。即使在天体运动中也存在着不可逆的摩擦力和热效应影响。因此，地球的旋转会由于潮汐的摩擦而逐渐减慢，随之而来的是月球逐渐地远离地球；若地球是一个完全刚性的旋转球体，就不会发生这种情况。

然而，事实仍然是，“物理的时钟式运转”仍然清晰地显示了非常突出的“有序来自有序”的那些特征——引起了物理学家的兴奋，一旦他们发现了生命体的这些特征。两者似乎还是有一些共同之处。究竟这种共同之处是什么，而明显不同的又是什么，才使得有机体的情形变得新奇和前所未见，这还有待发现。

NERNST ' S THEOREM

When does a physical system—any kind of association atoms—display 'dynamical law' (in Planck's meaning) or 'clock-work features'? Quantum theory has a very short answer to this question, viz. at the absolute zero of temperature. As zero temperature is approached the molecular disorder ceases to have any bearing on physical events. This fact was, by the way, not discovered by theory, but by carefully investigating chemical reactions over a wide range of temperatures and extrapolating the results to zero temperature—which cannot actually be reached. This is Walther Nernst's famous 'Heat Theorem', which is sometimes, and not unduly, given the proud name of the 'Third Law of Thermodynamics' (the first being the energy principle, the second the entropy principle).

Quantum theory provides the rational foundation of Nernst's empirical law, and also enables us to estimate how closely a system must approach to the absolute zero in order to display an approximately 'dynamical' behaviour. What temperature is in any particular case already practically equivalent to zero?

Now you must not believe that this always has to be a very low temperature. Indeed, Nernst's discovery was induced by the fact that even at room temperature entropy plays a astonishingly insignificant role in many chemical reactions. (Let me recall that entropy is a direct measure of molecular disorder, viz. its logarithm.)

能斯特定律

在什么时候，一个物理系统 —— 任何一种原子集合体 —— 才会表现出“动力学定律”（普朗克的意思）或“时钟式运转的特征”呢？量子理论对此问题有一个非常简洁的回答，即在绝对零度之时。

随着温度接近绝对零度，其分子的无序性便不再对任何物理事件产生影响。顺便说一句，这一事实不是通过理论发现的，而是通过仔细研究在很大的温度范围内的化学反应，将结果外推到绝对零度时得出来的，现实中绝对零度是达不到的。这就是瓦尔特·能斯特著名的“热定理”，有的时候也不过分地被骄傲地冠名为“热力学第三定律”（第一定律为能量原理，第二定律为熵原理）。

量子理论为能斯特的经验定律提供了理性基础，还使我们能够估算，一个系统要在多大程度上接近绝对零度，才会大致表现出“动力学的”行为。在任何一种特定情况下，什么温度实际上等同于绝对零度呢？

现在，你千万不要以为这一定是很低的温度。实际上，能斯特的发现正是根据这一事实推导得出的：即使在室温下，熵在许多化学反应中也起着令人惊讶的极其微不足道的作用（让我强调一下，熵是对分子无序性的直接度量，即它的对数）。

THE PENDULUM CLOCK IS VIRTUALLY AT ZERO TEMPERATURE

What about a pendulum clock? For a pendulum clock room 85 temperature is practically equivalent to zero. That is the reason why it works 'dynamically'. It will continue to work as it does if you cool it (provided that you have removed all traces of oil!). But it does not continue to work if you heat it above room temperature, for it will eventually melt.

THE RELATION BETWEEN CLOCKWORK AND ORGANISM

That seems very trivial but it does, I think, hit the cardinal point. Clockworks are capable of functioning 'dynamically', because they are built of solids, which are kept in shape by London-Heitler forces, strong enough to elude the disorderly tendency of heat motion at ordinary temperature.

Now, I think, few words more are needed to disclose the point of resemblance between a clockwork and an organism. It is simply and solely that the latter also hinges upon a solid—the aperiodic crystal forming the hereditary substance, largely withdrawn from the disorder of heat motion. But please do not accuse me of calling the chromosome fibres just the 'cogs of the organic machine'—at least not without a reference to the profound physical theories on which the simile is based.

摆钟几近绝对零度

那么摆钟的情况如何呢？对摆钟来说，室温实际上等同于绝对零度。这也是为什么它在“动力学”上行得通的原因。如果冷却它（只要你能去除所有的油痕），那它仍然会一如既往地不断运转下去。但是如果把它加热到室温以上，它就不会继续运转，因为它最终会熔化。

生命体与钟表装置

这看起来似乎无关紧要，但我认为它确实反映了问题所在。钟表之所以能够以“动力学”方式运转，是因为它是由固体构成的，这些固体在伦敦-海特勒力的作用下维持一定的形状，这种力的强度足以避免常温下热运动的无序倾向。

现在，我想，有必要再说几句以揭示钟表同有机体之间的相似之处，简而言之，有机体依赖着一种固体——构成遗传物质的非周期性晶体，从而很大程度上避免了热运动的无序性。请不要责怪我把染色体纤丝称为“生命机器的齿轮”——至少在没有参考这个比喻所基于的深刻的物理理论的情况下是这样的。

For, indeed, it needs still less rhetoric to recall the fundamental difference between the two and to justify the epithets novel and unprecedented in the biological case.

The most striking features are: first, the curious distribution of the cogs in a many-celled organism, for which I may refer to the somewhat poetical description on p. 79; and secondly, the fact that the single cog is not of coarse human make, but is the finest masterpiece ever achieved along the lines of the Lord's quantum mechanics.

因为，事实上，不需要太多的修辞就能直截了当地说明两者之间的基本区别，并证明在生物学的案例中，这些新颖且前所未有的绰号是正确的。

最显著的特点是：首先，这种齿轮奇妙地分布在一个多细胞有机体里，关于这点请参阅我在本章第4节（原文第79页）中作过的诗一般的比喻；其次，这种单个的齿轮不是粗糙的人工制品，而是按照上帝的量子力学完成的最为精致的杰作。

EPILOGUE
On Determinism and Free Will

86 As a reward for the serious trouble I have taken to expound the purely scientific aspects of our problem *sine ira et studio*, I beg leave to add my own, necessarily subjective, view of the philosophical implications.

According to the evidence put forward in the preceding pages the space-time events in the body of a living being which correspond to the activity of its mind, to its self-conscious or any other actions, are (considering also their complex structure and the accepted statistical explanation of physico-chemistry) if not strictly deterministic at any rate statistico-deterministic. To the physicist I wish to emphasize that in my opinion, and contrary to the opinion upheld in some quarters, *quantum indeterminacy* plays no biologically relevant role in them, except perhaps by enhancing their purely accidental character in such events as meiosis, natural and X-ray-induced mutation and so on—and this is in any case obvious and well recognized.

For the sake of argument, let me regard this as a fact, as I believe every unbiased biologist would, if there were not the well-known, unpleasant feeling about 'declaring oneself to be a pure

结语：决定论和自由意志

我费了很大的功夫，从纯科学的角度来阐述我们的问题，不怒不喜地（*sina ira et studio*, without anger and passion，既不发怒，也不热衷），作为奖励，请允许我加上我自己的，必然是主观的，对这个问题的哲学含义的看法。

根据前面的章节里提出的证据，一个生命体内发生的时-空事件，都与它的心灵活动，以及自我意识或其他活动相呼应，它们（还要考虑到这个生命体的复杂结构，以及我们可以接受的物理化学的统计学解释）即使不是严格意义上的决定论，至少也是统计决定论。对于物理学家，我想强调的是，在我看来，与某些方面得到支持的观点相左，量子不确定性在这些方面并不发挥任何与生物学相关的作用，也许诸如在减数分裂、自然发生的或X射线诱发的突变等事件中增强它们的纯粹偶然性方面除外，这在任何情况下都是明显和公认的。

为方便讨论起见，让我把它也看成事实，因为我相信任何一个没有偏见的生物学家都会那样做，如果他没有众所周知的

mechanism'. For it is deemed to contradict Free Will as warranted by direct introspection（内省）.

But immediate experiences in themselves, however various and disparate they be, are logically incapable of contradicting each other. So let us see whether we cannot draw the correct, non-contradictory conclusion from the following two premises:

(i) My body functions as a pure mechanism according to the Laws of Nature.

87 (ii) Yet I know, by incontrovertible direct experience, that I am directing its motions, of which I foresee the effects, that may be fateful and all-important, in which case I feel and take full responsibility for them.

不那么高兴的感觉，来“声称自己属于纯粹的机械论者”。因为这肯定是和直接发自内心的自由意志格格不入的。

但是，就直接经验本身来说，不管如何千差万别，千变万化，它们在逻辑上并不相互矛盾。那么，让我们来看看，从下述两个前提，我们能否得出正确的没有矛盾的结论：

(1) 根据自然规律，我的躯体是一套纯粹的机械装置；

(2) 然而我自己知道，根据无可争辩的直接经验，是我，在指挥我所有的行为，我可预见这些后果，这些后果可能是命中注定和至关重要的，在这种情况下，我会感受到并会对我所有的行为及其结果负全部责任。

The only possible inference from these two facts is, I think, that I—I in the widest meaning of the word, that is to say, every conscious mind that has ever said or felt 'I'—am the person, if any, who controls the 'motion of the atoms' according to the Laws of Nature.

Within a cultural milieu（社会背景）(*Kulturkreis*) where certain conceptions (which once had or still have a wider meaning amongst other peoples) have been limited and specialized, it is daring to give to this conclusion the simple wording that it requires. In Christian terminology to say: 'Hence I am God Almighty' sounds both blasphemous（亵渎神明的）and lunatic（疯狂的）. But please disregard these connotations for the moment and consider whether the above inference is not the closest a biologist can get to proving God and immortality at one stroke.

In itself, the insight is not new. The earliest records to my knowledge date back some 2,500 years or more. From the early great Upanishads the recognition ATHMAN = BRAHMAN (the personal self equals the omnipresent, all-comprehending eternal self) was in Indian thought considered, far from being blasphemous, to represent the quintessence of deepest insight into the happenings of the world. The striving of all the scholars of Vedanta was, after having learnt to pronounce with their lips, really to assimilate in their minds this grandest of all thoughts.

Again, the mystics of many centuries, independently, yet in perfect harmony with each other (somewhat like the particles in an ideal gas) have described, each of them, the unique experience of his

从这两个事实中唯一可能的推论，我想，是我——最广义上的“我”，也就是说，每一个曾经说过或感觉过“我”的有意识的头脑——如果存在，就是那个遵循自然规律控制“原子运动”的人。

在一个文化环境（*Kulturkreis*）之中，某些概念（在其他民族中，都曾经有过或仍然有着更为广泛的含义）都是有限定的并且有特定含义的，所以用结论所需的简单词汇去做这样的结论是大胆的。用基督教的话去说：“因此我是万能的上帝。”这听起来真是既亵渎又狂妄。在此时，请暂时忽略这些用词的含义，并考虑一下上述推理是否是一个生物学家所能够给出的最接近于证明上帝和永恒的推论。

就本身来说，这一见解并不新颖。据我所知，这一见解的最早记录可以上溯到2500年前以至更早。从古老而伟大的《奥义书》(Upanishads)中，自我（ATHMAN）等于梵（BRAHMAN）（个人自我正是无所不在，无所不包的永恒自我）在印度的思想中，并没有亵渎之意，代表了对世间万事万物的最深遂的真知灼见。所有的吠檀多（Vedanfa）学者所做的努力是，在学会用嘴唇发音之后，都真正地把这一伟大思想的精髓融入他们的心灵之中。

同样，许多世纪以来的神秘主义者，相互之间并没有联系，却又彼此和谐（有点像理想气体中的粒子），每个人描述他或她生命中的独特经历，都可以浓缩成一句：我已成为上帝（DEUS

or her life in terms that can be condensed in the phrase: DEUS FACTUS SUM (I have become God).

To Western ideology the thought has remained a stranger, in spite of Schopenhauer and others who stood for it and in spite of those true lovers who, as they look into each other's eyes, become aware that their thought and their joy are *numerically* one—not 88 merely similar or identical; but they, as a rule, are emotionally too busy to indulge in clear thinking, which respect they very much resemble the mystic.

FACTUS SUM）。

就西方的意识形态来说，这一思想仍然是陌生的，尽管叔本华等人全力支持，尽管那些真正的爱好者相互对眸时会意识到他们的所想所说已融为一体（用数字表示为1），不仅仅是相似或相同；但是他们通常在情感上过于忙碌而无暇专注于清晰的思考。在这一方面，他们与神秘主义者非常相似。

Allow me a few further comments. Consciousness is never experienced in the plural, only in the singular. Even in the pathological cases of split consciousness or double personality the two persons alternate, they are never manifest simultaneously. In a dream we do perform several characters at the same time, but not indiscriminately: we *are* one of them; in him we act and speak directly, while we often eagerly await the answer or response of another person, unaware of the fact that it is we who control his movements and his speech just as much as our own.

How does the idea of plurality (so emphatically opposed by the Upanishad writers) arise at all? Consciousness finds itself intimately connected with, and dependent on, the physical state of a limited region of matter, the body. (Consider the changes of mind during the development of the body, as puberty, ageing, dotage, etc., or consider the effects of fever, intoxication, narcosis, lesion of the brain and so on.) Now, there is a great plurality of similar bodies. Hence the pluralization of consciousnesses or minds seems a very suggestive hypothesis. Probably all simple, ingenuous people, as well as the great majority of Western philosophers, have accepted it.

It leads almost immediately to the invention of souls, as many as there are bodies, and to the question whether they are mortal as the body is or whether they are immortal and capable of existing by themselves. The former alternative is distasteful while the latter frankly forgets, ignores or disowns the facts upon which the plurality hypothesis rests. Much sillier questions have been asked: Do an-

请允许我再说那么几句，意识永远不会以复数（多元）形式出现，其只会以单数（一元）形式出现。即使在意识分裂或双重人格的病态情况下，两种情况也是交替出现，从来不会同时表现出来。在梦境之中，我们确实同时扮演几个角色，但并不是没有区分：我们扮演其中一个角色，在这个角色身上，我们直接行动和说话，同时常常焦急地等待另一个角色的回答或回应，却没有意识到，是我们自己在控制这个角色的言行，正如控制着我们自己的言行一样。

这个多元化（这是《奥义书》的作者们显然不能接受的）的概念是如何产生的呢？意识发现自己与物质的有限区域——身体——的物理状态密切相关，甚至依赖它（考虑我们躯体发育的过程中心灵的变化，如青春发育、成年、衰老等；或考虑发烧、中毒、麻醉、脑损伤等的影响）。现在，有那么多元的相似躯体。因此，意识或思想的多元，似乎是一个很有启发性的假设。或许所有单纯、质朴的人们，还有绝大多数西方哲学家都接受了它。

这就几乎立刻引出了灵魂的发明，有多少躯体，就有多少灵魂，引出了灵魂是像肉体一样会死还是灵魂是不朽的，能够自我存在的问题。前一个问题显然不讨人喜欢，而后者则坦率地忘记、忽视或否认了多元化假说所依据的事实。还有很多愚蠢的问题也被提出来了，动物也有灵魂吗？甚至有人质疑女性是否有灵魂，是否只有男性才有灵魂。

imals also have souls? It has even been questioned whether women, or only men, have souls.

Such consequences, even if only tentative, must make us suspicious of the plurality hypothesis, which is common to all official Western creeds. Are we not inclining to much greater nonsense, if in discarding their gross superstitions we retain their naïve idea of plurality of souls, but ‘remedy’ it by declaring the souls to be perishable, to be annihilated with the respective bodies?

这样的结果，即使只是暂时的，也一定能使我们怀疑多元化这个所有西方的官方都遵循的教义假说。如果我们不认可这些胡说，而保留他们那些天真的灵魂多元化的观点，但又通过声称灵魂是可以随着新附躯体的毁灭而消亡的观点去“补救”它，难道不是更大的谬误吗?

89 The only possible alternative is simply to keep to the immediate experience that consciousness is a singular of which the plural is unknown; that there *is* only one thing and that what seems to be a plurality is merely a series of different aspects of this one thing, produced by a deception (the Indian MAJA); the same illusion is produced in a gallery of mirrors, and in the same way Gaurisankar and Mt Everest turned out to be the same peak seen from different valleys.

There are, of course, elaborate ghost-stories fixed in our minds to hamper our acceptance of such simple recognition. E.g. it has been said that there is a tree there outside my window but I do not really see the tree. By some cunning device of which only the initial, relatively simple steps are itself explored, the real tree throws an image of itself into my physical consciousness, and that is what I perceive. If you stand by my side and look at the same tree, the latter manages to throw an image into your soul as well. I see my tree and you see yours (remarkably like mine), and what the tree in itself is we do not know. For this extravagance Kant is responsible. In the order of ideas which regards consciousness as a *singulare tantum* it is conveniently replaced by the statement that there is obviously only *one* tree and all the image business is a ghost-story.

Yet each of us has the indisputable impression that the sum total of his own experience and memory forms a unit, quite distinct from that of any other person. He refers to it as 'I' and *What is this* 'I'?

唯一可能的选择也许是简单地坚持直接经验。这就意味着意识是一元形式的，其多元性仍是未知；只有一种事物，这种看似多元的事物，也不过是这一事物的不同的方面，由一种欺骗（印度之MAJA）所产生；同样的错觉产生于镜子画廊，同样，高里三喀峰（Gaurisankar）和珠穆朗玛峰其实只是从不同山谷中看到的同一座山峰而已。

当然，那些精心编造的鬼怪故事，已经固定在我们的脑子里，有碍我们对那些简单的认识的接受。例如，有人曾说我的窗外有那么一棵树，但我并没有真正看到那棵树。借用一些精巧的装置，尽管还只是最初的相对简单的几步尝试，那棵实实在在的树的影像就投射在我的物理意识之中，而这正是我所感知的东西。如果你也站在我的身边，也来看这同一棵树，这棵树也能把这一影像投射到你的意识之中。我看到了我的树，你看到了你的树（当然极像我的那棵树），而这棵树本身是什么我们并不知道。对这一出格的说法，康德是有责任的。意识是一元形式的这种观念，很容易用这样的陈述取代：显然那里只有一棵树，所有影像只不过是鬼怪之类的怪诞之说。

然而，我们每一个人都有自己的一种根深蒂固的印象，即自己的经验和记忆的总和形成了一个与任何别人的印象截然不同的整体，他把它称为“我”，这个“我”又是什么呢？

If you analyse it closely you will, I think, find that it is just a little bit more than a collection of single data (experiences and memories), namely the canvas *upon which* they are collected. And you will, on close introspection, find that what you really mean by 'I' is that ground-stuff upon which they are collected. You may come to a distant country, lose sight of all your friends, may all but forget them; you acquire new friends, you share life with them as intensely as you ever did with your old ones. Less and less important will become the fact that, while living your new life, you still recollect the old one. 'The youth that was I', you may come to speak of him in the third person, indeed the protagonist of the novel you are reading 90 is probably nearer to your heart, certainly more intensely alive and better known to you. Yet there has been no intermediate break, no death. And even if a skilled hypnotist（催眠师）succeeded in blotting out entirely all your earlier reminiscences（回忆录）, you would not find that he had killed *you*. In no case is there a loss of personal existence to deplore.

Nor will there ever be.

如果你再仔细分析一下，我想，你就会发现这些事实只不过是一些单一的数据（经验和记忆）的集合，也就是一张集合了这些经验和记忆的画布。而你只要仔细反思，就会发现你所说的“我”真正的含义就是收集经验和记忆的基础物质。你可能来到一个远方的国度，与你所有的朋友都失去了联系，甚至几乎将他们遗忘；你结识了新朋友，你与他们一起分享生活，如同曾经的老朋友一样。

当你在经历这样的新生活的时候，尽管事实上过去的一切变得越来越不那么重要，而你仍会回忆那过去的时光。“那个曾经是我的青年”，你也许会用第三人称来称呼他，的确，你正在阅读的那本小说中的主人翁也许与你的心贴得更近，更加栩栩如生，知之甚多。然而现在的你和过去的你也没有经历过什么生离死别。即使一个经验丰富的催眠大师，成功地从你的脑海中抹去早先的所有记忆，你也不会发觉他已经终结了那个“你”。在任何情况下，你不会去为你那失去的自我存在而痛心疾首。

现在不会，将来也永远不会。

NOTE TO THE EPILOGUE

The point of view taken here levels with what Aldous Huxley has recently—and very appropriately—called *The Perennial Philosophy*. His beautiful book (London, Chatto and Windus, 1946) is singularly fit to explain not only the state of affairs, but also why it is so difficult to grasp and so liable to meet with opposition.

结语之注释

我在这里陈述的观点，与阿尔道斯·赫胥黎最近发表的——非常合适地——称之为《长青哲学》（查托和温德斯出版社，伦敦，1946）中的思想相当一致，这一美妙之作不仅最为合适地解释这一状况，而且还特别适合于解释为什么这些观点是如此难以理解而又如此容易遭到反对。

编译说明

3年前，在全球欢庆国际人类基因组计划协作组宣布人类全基因组序列草图完成二十周年之际，我们欣然接受了湖南科学技术出版社首席编辑的邀请，于该年5月27日接受了将诺贝尔物理学奖获得者薛定谔的《生命是什么？》再度翻译成中文的艰巨任务。借此机会，先向读者谈谈本书编译的体会。

我接受这一任务的正面理由和动机有三：

一、几位诺奖得主的推荐：我曾有机会与所认识并有交往的几位诺贝尔生理学或医学奖得主畅谈。正像平常一样，我这位被昵称为“提问专家（Question-Asking Expert）者”，在向他们请教了很多问题之后，从来没有忘记再问最后一个问题：“对您印象最深、影响最大的是哪一本书？”有趣的是，他们无一例外、不约而同地说：“薛定谔的《生命是什么？》。”

我去年再次造访圣三一学院之时，在薛定谔1943年2月做报告的阶梯教室门口，看到Watson和Crick给他的信件，和他在自己的几本书中所写的一样，声称他俩发现DNA双螺旋结构是受薛定谔这本书的影响。我当然不相信他们这样的人物也是人云亦云。看来此书常被称为“世纪巨著的世纪之问”不是没有道理的。

二、与薛定谔的“隔世缘分”：我曾在20世纪初，应时任爱尔兰圣三一学院生命科学院院长D. McConnell邀请，同一天在他的学院做了两场报告。第一场是学术专场，有关我们中国对人类基因组计划的贡献等；第二场是与大学生交流即近似于科普性质的大讲座。主人告知，薛定谔当年做报告的阶梯教室就在对面的物理学大楼里。遗憾的是，那次没能立即去瞻仰这一历史性的场地，去年才特别地补上这一遗憾，自然又有一番感受和联想。

三、翻译是最好的阅读：我真地好几次读过这本书，尽管没有像读另外一些书那样精读（我对读书的标准之一是做读书笔记），但绝不只是我自己定义的“泛读”。我真的想借翻译之机再精读这本书。不能说是精读，至少可以说逐章逐段逐句逐词逐字逐标点细读几遍。可以说三年所得，首先是“精读”此书的个人收获。

这三点是正面的，而负面的理由则是一场挑战。

我是写过书，也译过书的，深知写书是“软伤”，错的是自己的思想（自有其说），译书则是“硬伤”，暴露的是自己的肤浅或无知（铁证如山）。

我之所以将此书的再次翻译看成是自己在写书与译书生涯中最严峻的挑战，是因为这本书自2003以来，我国至少有多家出版社，10多位汉语译者出版了近20个版本（湖南科学技术出版社就出版/再版了4个版本）。其中一定不乏物理学科班出身者和专业翻译人士。可以肯定地说，这些译者在英汉互译技巧和专业（生物学、物理学、翻译学等）上各有千秋，在“信、达、雅”三方面都很有成就。但不可否认的是，也有一些常识性的错误（如把“at the beginning of the 1950s”译成“1950年初”，把Smithsonian等同于Smith等）。也许那些译者或编辑认为这些问题并不会影响读者对原文的阅读及对原义的理解，但我们能苟同吗?

我一定要向所有的读者直率地承认，我的这个版本相当一部分是取人之长，择优润色；另一部分是自己从头译起，不参照任何别的译本。就翻译文字来说，恐怕不可能与以前的版本完全没有直接关联，当然也没有多大“超越”。借此，我要向所有版本的译者，包括同学和同事表示衷心的感谢。如果这本书有什么精彩之处，都应理所当然地归功于他们。如果这版本仍有错误，那我将更为歉疚羞愧，因为我的专业和英文水平真的见底了！离“信、达、雅”甚远：求“信”而成“直译”，求“雅”而偏离本意，只能改“雅”为“顺”，仅求顺畅了。当然，最担心的还是“复制”他人的错误，这常被作为“照抄”的证据。

本译文在刚开始准备英汉对照时，原意是让读者去比较、来鉴别，只是在偶数页上附上对应的英文原文，并使中文翻译语序尽量接近英文原文，后又参照一些艺术经典的英汉对照本，对少数自认为较少见而又较重要的英文词汇也加上汉注。客观上，正如一位好友之告诫——这是“自取其辱”，兜出了自己的英语水平以至于词汇量。我作为一个教书匠，近几年，一直“别出心裁”地在讲课或讲座的第一张PPT上，大言不惭地写上“纠一错，赠一书”。秉承“知为知，不知则不知”的为师法则，我在认错时，向来是“面不改色心照跳”并一定立即改正，但也不否认可能会被人认为是“狂妄”。这里我倒真心地希望本译作的读者尽管放心大胆地提出批评，以求“译与读互动”，在再版或再印时，通过译者与读者的共同努力，薛定谔这一世纪巨著的汉译本能精益求精，更上一层楼。

有一点倒是真的考虑到了：2023年2月是本书的思想源泉——薛定谔先生在圣三一学院讲座80周年，也恰逢国际人类基因组计划落下帷幕20周年。我们将在今年4月份，作为国际基因组学大会第18届年会（ICG-18）的主题之一，举行国际人类基因组计划完成20周年，生命科学的第一语录——杜布赞斯基有关演化的一句话发表50周年，DNA双螺旋结构发现70周年，特别是《生命是什么？》的系列报告发表80周年的学术讨论会，届时将邀请爱尔兰的朋友欢聚一堂，探讨这一历史巨著对人类知识宝库的贡献和各自的读后感，交换对“生命是什么”这一千古命题的更多更新的探索。

有幸的是，我发现这本书的法文版译者居然是法兰西院士、我的良师益友安东·唐善（Antoine Danchin）教授。自从我在21

世纪初认识他以来，我一直对他十分佩服。举个例子，他每周都会在网上举行一次全球性研讨会（Seminar），20多年来从未间断。我从未见过其他人像他一样，每周都能提出一个如此精彩的题目，每次2至3页的英文内容，都是他自己打字打出来的。唐善教授本是物理学科班出身，同时也是享有国际盛誉的分子生物学和生物信息学专家。他欣然应邀为我们的中文版作序，洋洋洒洒10余页，他真是我见过的对薛定谔这本小书内容和观点十分熟悉，对它的肯定和批评最为中肯的人。他的序确为本书锦上添花。这也体现了他对此书的钟爱，阅读的仔细和理解的全面。他在讨论这些问题时旁征博引，也反映了他学识渊博，这也是我们邀请他为本书撰写这一长长的序言的另一目的。

还有几点：(1) 本书对英文原文不做任何加注或纠正那些明显的错误（唯一的例外是1956年华裔遗传学家蒋有兴和瑞典遗传学家阿尔伯特·莱万已证明人类有46条染色体）；(2) 本文正文后面特设较为详尽的“检索”，特别是专业术语的英中对照，尽管不可能完整；(3) 特设人名的英中对照索引以便大家查询。

当年，这本书的出版为很多前辈带来了多方面的启示。今天又有很多新的生物物理学上的突破：从最初的X射线衍射图谱推衍出DNA螺旋结构，到获2017年诺贝尔化学奖的冷冻电镜技术阐明蛋白质三维结构的重大贡献，再到今天的蛋白质结构的AI预测，也都可在一定程度上归功于这本书的思想与思考的贡献。我想我们这几代人，也许都应该重温《生命是什么？》，更深入地体会这本书是如何启迪了我们生命科学的前辈，并吸引诸多物理、化学、数学及其他学科的吾辈或晚辈投身到解码生命中来。正如本书原序所言:“为理解此书‘持续至今的重要意义，它非常值得我们再次阅读’。”

最后，谨向参与这一工作并做出重大贡献的周润、王元梅、姜环、张馨文、王晓玲和董伟等同事，特别是哥本哈根大学校友、中国科学院理论物理研究所喻明教授所作的物理专业校准与全文修饰润色，以及湖南科学技术出版社首席编辑孙桂均、责编李蓓、金振坤及其他同事为此书能在80周年与读者见面所作的努力，表示衷心的感谢。

杨焕明

2023年3月17日

人名的英中对照索引

续表1

续表2

部分专业术语的英中对照索引

续表1

续表2

续表3

英文	中文	页码
deviation	偏差	12,38,88
dichromasy	双色色盲	84
diffusion	扩散	30–34,38,232
diploid	二倍体	62–66
dipole axes	偶极轴	26
directed mutations	定向突变	94
discrete	离散的	76,134–138
disintegration	衰变	146,232
dominance	显性	72,94
driving mechanism	驱动机制	246
egg cell	卵细胞	52–54,60–62,66,84
electric motor	电动机	224
electric pulses	电脉冲	242
electrometer	静电计	244
electron-volt	电子伏特	182,190
energy level	能级	90,134–136,194
energy threshold	能量阈值	150,156
entelechy	活力	208
entropy	熵	196,202–204,210–222,236,250
evolution	演化，进化	94,184

续表4

续表5

英文	中文	页码
Heitler-London forces	海特勒–伦敦力	172,252
hereditary substance	遗传物质	90–92,156,162,196–200,252
hermaphroditic	雌雄同体的	100
heterocyclic ring	杂合闭环	82
heterozygosy	杂合子	72
heterozygous	杂合的	94,100–102,106,110
homologous chromosomes	同源染色体	70–74,96,116
homozygous	纯合的	100–106,110
inanimate matter	非生命物质	228,236
inert lump of matter	惰性物质	204
inert state	惰性状态	208,216
ionization	电离	120–122,190
irregular heat motion	不规则热运动	216
isomeric	同分异构的	146,150–152,156–160,176,182,190
isomerism	同分异构	150,160,176
large-scale hereditary characteristics	宏观遗传性状	228
large-scale mechanical phenomena	宏观的机械现象	242
lattice	晶格	126,166,214

续表6

续表7

续表8

续表9

英文	中文	页码
sterilization	绝育	124
syngamy	配子配合，融合生殖	60–62
The Dynamical and the Statistical Type of Law	动力学和统计学定律	242
thermodynamical	热力学的	202,220
thermodynamical equilibrium	热力学平衡	204,218
Third Law of Thermodynamics	热力学第三定律	250
torsional balance	扭力天平	40,244
universal attraction	万有引力	240
vapour pressure	蒸气压	206
vertical axis	垂直轴	40
viscosity	黏度，黏性	30,168
wave-length	波长	12,120,190
wave mechanics	波动力学	132
X-ray	X 射线	116,120,124–128,168,184,190–192,256
yard	码	14
γ-ray	γ 射线	116,120

图书在版编目（CIP）数据

生命是什么：汉英对照 /（奥）埃尔温・薛定谔著；杨焕明译．—长沙：湖南科学技术出版社，2023.4
书名原文：What is life?
ISBN 978-7-5710-2111-5

Ⅰ．①生…　Ⅱ．①埃…　②杨…　Ⅲ．①生命科学—普及读物—汉、英　Ⅳ．① Q1-0

中国国家版本馆 CIP 数据核字（2023）第 049671 号

SHENGMING SHI SHENME
生命是什么

著者
[奥] 埃尔温・薛定谔
译者
杨焕明
出版人
潘晓山
策划编辑
孙桂均
责任编辑
李蓓　金振坤
营销编辑
周洋
出版发行
湖南科学技术出版社
社址
长沙市芙蓉中路一段 416 号
泊富国际金融中心
网址
http://www.hnstp.com
湖南科学技术出版社
天猫旗舰店网址
http://hnkjcbs.tmall.com

印刷
长沙超峰印刷有限公司
厂址
宁乡市金州新区泉洲北路 100 号
邮编
410600
版次
2023 年 4 月第 1 版
印次
2023 年 4 月第 1 次印刷
开本
880 mm × 1230 mm　1/32
印张
10.25
字数
222 千字
书号
ISBN 978-7-5710-2111-5
定价
58.00 元